"Tavinor's earlier work on the aesthetics of video games kick-started philosophical research in that area; this book will do the same for the aesthetics of virtual reality."

— *Nathan Wildman, Tilburg University*

The Aesthetics of Virtual Reality

This is the first book to present an aesthetics of virtual reality media. It situates virtual reality media in terms of the philosophy of the arts, comparing them to more familiar media such as painting, film and photography.

When philosophers have approached virtual reality, they have almost always done so through the lens of metaphysics, asking questions about the reality of virtual items and worlds, about the value of such things, and indeed, about how they may reshape our understanding of the "real" world. Grant Tavinor finds that approach to be fundamentally mistaken, and that to really account for virtual reality, we must focus on the medium and its uses, and not the hypothetical and speculative instances that are typically the focus of earlier works. He also argues that much of the cultural and metaphysical hype around virtual reality is undeserved. But this does not mean that virtual reality is illusory or uninteresting; on the contrary, it is significant for the altogether different reason that it overturns much of our understanding of how representational media can function and what we can use them to achieve.

The Aesthetics of Virtual Reality will be of interest to scholars and advanced students working in aesthetics, philosophy of art, philosophy of technology, metaphysics, and game studies.

Grant Tavinor is Senior Lecturer in Philosophy at Lincoln University, New Zealand. He has published widely on the aesthetics of videogames, virtual worlds, digital media ethics, and the philosophy of technology.

Routledge Research in Aesthetics

Philosophy and Film
Bridging Divides
Edited by Christina Rawls, Diana Neiva, and Steven S. Gouveia

Paintings and the Past
Philosophy, History, Art
Ivan Gaskell

Portraits and Philosophy
Edited by Hans Maes

Radically Rethinking Copyright in the Arts
A Philosophical Approach
James O. Young

Philosophy of Sculpture
Historical Problems, Contemporary Approaches
Edited by Kristin Gjesdal, Fred Rush, and Ingvild Torsen

Art, Representation, and Make-Believe
Essays on the Philosophy of Kendall L. Walton
Edited by Sonia Sedivy

Philosophy of Improvisation
Interdisciplinary Perspectives on Theory and Practice
Edited by Susanne Ravn, Simon Høffding, and James McGuirk

The Aesthetics of Virtual Reality
Grant Tavinor

For more information about this series, please visit: https://www.routle-dge.com/Routledge-Research-in-Aesthetics/book-series/RRA

The Aesthetics of Virtual Reality

Grant Tavinor

Routledge
Taylor & Francis Group

NEW YORK AND LONDON

First published 2022
by Routledge
605 Third Avenue, New York, NY 10158

and by Routledge
2 Park Square, Milton Park, Abingdon, Oxon OX14 4RN

Routledge is an imprint of the Taylor & Francis Group, an informa business

© 2022 Grant Tavinor

Library of Congress Cataloging-in-Publication Data
Names: Tavinor, Grant, author.
Title: The aesthetics of virtual reality / Grant Tavinor.
Description: New York, NY : Routledge, 2022. | Series: Routledge research in aesthetics | Includes bibliographical references and index.
Identifiers: LCCN 2021018485 (print) | LCCN 2021018486 (ebook) |
ISBN 9780367619251 (hbk) | ISBN 9780367620424 (pbk) |
ISBN 9781003107644 (ebk)
Subjects: LCSH: Virtual reality--Psychological aspects. | Human-compuer interaction--Psychological aspects. | Digital media--Psychological aspects. | Interactive multimedia--Psychological aspects. | Aesthetics.
Classification: LCC QA76.9.H85 T38 2022 (print) | LCC QA76.9.H85 (ebook) | DDC 006.8--dc23
LC record available at https://lccn.loc.gov/2021018485
LC ebook record available at https://lccn.loc.gov/2021018486

ISBN: 978-0-367-61925-1 (hbk)
ISBN: 978-0-367-62042-4 (pbk)
ISBN: 978-1-003-10764-4 (ebk)

DOI: 10.4324/9781003107644

Typeset in Sabon
by Taylor & Francis Books

For my wonderful wife Dorothy

Contents

Acknowledgements

This book grew out of my earlier work on gaming, particularly a chapter in *The Aesthetics of Videogames*, a collection I edited with Jon Robson also for the Routledge Research in Aesthetics series. That chapter explored, in a very provisional way, some of the issues that have found their way into the following pages. Like that chapter, I see the current book as modest and exploratory, in setting out an approach grounded in the method and theory of the philosophy of the arts, and also paying careful attention to the technological and cultural phenomenon of virtual reality (VR) itself. Foremost here, I wanted to contextualize virtual reality within the history and practice of picture making and aesthetics, in a way that would allow this book to escape the metaphysical confusions so typical of other academic approaches to this topic. Given the gravity of those metaphysical constellations, this this was not an easy task!

This book is about a picturing phenomenon, but the reader will note that it contains no pictures whatsoever. This is intentional, as the reproduction of the egocentric and interactive pictorial form theorized in this book would be impossible in the picturing format available here. A small black and white still image can hardly give much of an impression of the pictorial phenomena I discuss in this book. Nevertheless, I have been careful to supply the date of release of the VR games and applications referred to. I envisage the reader actually playing through or experiencing some of the examples as they read the book (though perhaps not simultaneously). This may help readers keep themselves virtually grounded in the phenomena under discussion.

I would like to thank Andrew Weckenmann and Allie Simmons, and everybody at Routledge who has shown confidence in this project and allowed me the freedom to undertake the exploration of these issues.

My university department at Lincoln University in New Zealand has consistently allowed me the space to complete this research, and deserve thanks for this, and also for a small research fund that allowed for a short-term research assistant to chase up some of the more technical examples and literature concerning VR media.

The philosophers, too many to list, who have encouraged me or commented on my work also deserve thanks. Specific appreciation goes to audiences at the Pacific Division meeting of the American Philosophical Association meeting in

San Diego in 2018; the annual meeting of DiGRA [Digital Games Research Association] in Kyoto in 2019; the Pacific Division of the American Society for Aesthetics held at Berkeley in 2019; and at the 77th and 78th Annual meetings of that Society in 2019 and 2020 (the latter held virtually, during to the ongoing COVID-19 epidemic). The comments and questions at these sessions helped to sharpen the ideas here considerably. I would also like to thank my commentators for the American Society of Aesthetics conferences, Katherine Thomson-Jones, Stephanie Patridge, and Michalle Gal, whose comments and critiques also improved this work.

Finally, I would like to thank my wife Dorothy Cabunilas Tavinor for her companionship, love, and forbearance during the long hours I sat at the kitchen table writing this book during an eventful 2020.

Christchurch, March 2021

1 The Virtual Turn

1.1 An Introduction to Virtual Reality

Once I had everything set up, with the numerous cords and connections properly plugged in and a space cleared in front of me so I would not trip over, I turned on the console, entered the disc, and put on the headset.

Oh, this is what virtual reality is like.

Of course, I knew what virtual reality was in an abstract sense—it presents visual scenes that give the impression that you occupy an alternative place—but the first-hand experience felt a bit like a revelation. *It really does feel like I am within this place.* And then I turned my head to discover that the space also existed *behind me*. To discover that I could reveal the world by simply turning my head to see my surroundings, was another surprise. I could not stop looking around, taking pleasure in inspecting the apparent illusion that I occupied another place. And this was just the menu screen for the demo disc! Virtual reality (VR) gets a lot of hype, but some of it surely is deserved. Putting on the headset for the first time, the sensory impression of being in another space is immediate and undeniable.

Soon afterwards I received a copy of the survival horror game *Resident Evil VII* as a sort of sadistic gift from my brother. *Resident Evil* is a series of videogames in which the player finds themselves trapped and helpless in some mysterious and threatening scenario. There are zombies, oversized rabid dogs, that kind of thing. I had played *Resident Evil* games and others from its genre previously, but while challenging and frustrating, I never found them particularly scary. In such cases it is all too easy to look away from the screen and distance yourself from the horror. I found the VR version of *Resident Evil* to be an altogether different story. The game begins with the player entering an apparently abandoned and dark country house, slowly searching through the house to solve a mystery. But even the first hallway was utterly terrifying for me to experience. I stood in the entrance for a long time wondering what was ahead. I did not want to walk down the hallway knowing that I would have to open a door at its end, and who knew what was behind that door?!

DOI: 10.4324/9781003107644-1

As a result, my progress in the game was incredibly slow, because each time I entered a new area I would stop and carefully look and listen for threats in the environment. Every time I inspected an object in the world, I worried that because my attention was on the item some threat in the house would sneak up and do me some horrid evil. I also did a lot of stopping and looking over my shoulder to make sure I was not being followed. I felt vulnerable. When I finally did meet the inhabitants of the house, the involuntary terror that overtook me in that moment was so complete that I did something I cannot remember ever doing while playing a game before: *I literally screamed.* This must have all seemed hilarious to my wife who observed from the other side of the room, but she was also now quite reluctant to enter the virtual house herself. After that I had to take off the headset for a little while. It was a terrifying experience; I can thoroughly recommend it.

Virtual reality has been with us for at least 40 years, both as a preoccupation of technologists and futurists, and by existing as a viable though cumbersome and creaky technology usually designed for games or technical applications. Virtual reality has also had an undeniable influence on popular culture, particularly as a subject of films such as *Tron* (1982) or *The Matrix* (1999) and the Holodeck on the TV show *Star Trek: The Next Generation* (1987–1994). But it is only very recently that the medium itself has become widely available for home use. Several virtual reality products are now commercially available, including dedicated VR headsets from PlayStation 4, Valve, HTC and the virtual reality forerunner Oculus, and a number of mobile headsets that use smartphones as their screen, such as Samsung Gear VR. There are signs, perhaps, that virtual reality is not the fleeting novelty it could have been, and that many critics predicted it would be. And there are now many popular games that utilize the technology. *No Man's Sky* (2016) for example, has an often-beautiful virtual reality mode that allows players to explore galaxies of procedurally generated planets. The venerable videogame series *Half-Life* has received a VR sequel in the form of *Half-Life: Alyx* (2020). Because of its recent successes, it is increasingly common to hear that "Virtual reality is finally here." This may be premature, because there are still deep challenges to the medium and its adoption by a truly mass audience, but the practical realization of the concept already seems before us.

These events are, foremost, a *media development*. What are the key features of this advancement in media? I noted above that the basic understanding of virtual reality is that it "presents visual scenes as though you occupy a place within them," but in technical terms, what does this involve? Leaving aside a lot of detail and theory for the following chapters, in the most general terms, PlayStation VR and other commercial virtual reality systems comprise three key elements: the depiction of a sensory environment; a means of tracking and depicting the user's apparent position within this environment; and, finally, a means of providing for user interaction within this virtually depicted space. These, as I will refer to the phenomenon here, are the principal elements of *virtual reality media* (or *VR media* for short).

The depiction of the VR environment is predominantly visual and is most frequently achieved via a stereoscopic headset or head mounted display (HMD). For example, the PlayStation VR headset includes a single small organic light-emitting diode panel that sits in the headset close to the user's eyes. Two small lenses are placed before the screen, magnifying and softening the images, allowing for a wider field of vision on the depicted scene, and reducing the visual prominence of the surface of the pixel array. The images that are depicted on the panel are produced in such a way—and this is where the sophisticated rendering algorithms of the software play a role—that the apparent displacement of the two images, combined with the binocularity of the lenses, mimics our natural visual situation in the real world. The resulting visual environment—such as a darkened room within a decaying house—gives a very strong visual impression that the viewer is situated within that environment.

While the visual elements of VR media predominate, the depiction of the sensory environment is not restricted to this visual modality, and at a minimum VR media usually involve stereophonic, or even 3D virtual audio, to place the user within an aural space. The spatial effect is achieved in a formally similar manner to the stereoscopic depiction of the visual environment: the spatial cues of native hearing—the displacement of the two ears and the brain's ability to use the resulting difference in timing and intensity of the received sounds to identify the spatial location of sound sources, that is, the ability of *sound localization*—is utilized in VR to mimic the acoustic spatiality of natural hearing.

Sometimes a means of haptic or kinesthetic representation is also used to convey a sensory engagement with the virtual environment; the senses in question being touch, proprioception and the spatial sense provided by the vestibular system. Haptic gloves, which provide tactile feedback when you touch or grasp objects in the virtual environment, are now being developed by several companies. But despite these developments, these modes of sensory representation remain less common and quite limited compared to the visual and auditory modes introduced above.

This apparent experiential vantage point is not passive or fixed, as VR systems typically allow for the movement of the user to be tracked and for this movement to be represented in the user's apparent experiential orientation on the virtual environment. In PlayStation VR, this tracking partly involves small light emitting diodes (LEDs) fixed to the exterior of the headset which are captured by a camera placed in front of the user. The relative position of the LEDs is used to calculate the orientation of the user's head so that head movement can be replicated in the depictive viewpoint displayed by the stereoscopic headset. The HTC Vive goes further and allows for the user's body movements to be tracked and depicted within a 15-foot radius. Other VR systems such as Oculus and Magic Leap have developed eye tracking technology so that the experiential effects of eye movement—particularly on focus in the visual field—might be replicated. For each of these means of tracking to be effective, a low latency between the tracking and the display is crucial for creating a realistic impression, and this has been a significant technical hurdle in the design of effective VR systems.

While many VR media applications comprise "experiences" where the user's agency is limited to changing their orientation on the depicted virtual spaces, most applications, including all VR games, allow the user the ability to interact with objects within the virtual space. So, for example, in the opening scene of *Half-Life: Alyx*, players find themselves on a balcony looking over a dystopian city dominated by a vast Citadel, built by the Combine, the antagonist force of the game series. Combine craft move menacingly through the air and stalk the streets of City 17. On the balcony, the player can interact with various objects depicted in the world of the game: a radio can be tuned to various stations, cans of soda can be picked up, crushed, and thrown (perhaps at a pigeon perched on the balcony rail). In VR, such control is typically achieved either through standard gaming controllers or purpose-built peripherals, but again the position and orientation of these devices needs to be tracked by the VR system so that the user's movements and interactions can be realized in the virtual space.

Together, these media elements may give the users of VR a compelling sense of inhabiting and interacting with alternative spaces. And these alternative experiential spaces are somewhat more diverse than is frequently acknowledged. These media may give rise to the fear and trepidation that I felt while playing *Resident Evil VII*, but they may liberate us to explore fantasy spaces such as in *No Man's Sky* where the predominating feeling need not be one of vulnerability, but instead one of curiosity. VR media also allow us to explore and learn about the landscapes and cities of our *actual world* as in *Google Earth VR*, which utilizes the mapping technology with which we are now so familiar, to present navigable VR depictions of our world's geography. We may even step into historical spaces to experience ancient cities such as in the independent developer Steven Lou's *VR Rome* (2018). VR may also free us from our human-sized spatial constraints so that we can experience the very large and the very small, including the VR experience *Discovering Space* (2017) which allows users to explore the solar system, and the virtual microscopy VR app arivis VisionVR, in which users can manipulate and section microscopic items such as cells. And indeed, VR media may allow for the disruption of our conception of space, as in *Superliminal* (2020), a VR game that employs forced perspective and other visual devices to create challenging puzzles. Finally, VR may affect the spaces in which art takes place, as in the VR painting application *Tilt Brush*, which can be used to create paintings that because of their three-dimensional appearance seem like a kind of virtual sculpture.

1.2 The Aesthetics of Virtual Reality Media

These virtual reality media are a single aspect of a widespread techno-cultural trend of recent times. So many of the artefacts, activities, institutions, and relationships which in previous times we experienced in the actual world, we now encounter often, if not principally, in a *virtual* way. You might, like me, spend your leisure time playing virtual reality games or exploring virtual worlds. Virtual stores, a novelty just twenty-five years ago, are now ubiquitous features in our lives, and some of us will do most of our shopping through such online means. You may now interact with many of your "friends" in a principally virtual way by posting on social media or "liking" their posts and activities. Some of these virtual friends you may not have even encountered face to face in the actual world. Our economies are now in a significant part distributed in digital worlds, distant from the facts of labor, products, and services, with many items of trade seeming hardly tangible at all. Our politics has also become increasingly virtualized, in that our leaders may communicate with us through virtual means, utilizing the same social media we use to chat with our friends, to communicate with or influence their followers. We might refer to this trend as the *virtual turn*.

While acknowledging that my focus is part of a broader set of techno-cultural changes, and that a basic examination of the concept of virtuality needs to be given, I will for the most part avoid discussion of this wider context. Rather, I will address one specific aspect of the virtual turn: the way that virtual reality media allow their users the impression of experiencing environments other than those in which they actually exist. Moreover, my interest will be in the philosophical and aesthetic questions raised by such virtual media. The kinds of questions I ask, and issues I discuss, will derive from my disciplinary orientation as a philosopher of the arts.

The philosophy of the arts, or what is also frequently called philosophical aesthetics, is an active and robust subdiscipline of philosophy that enquires into the nature of the arts and aesthetic experience. In the last fifty years in particular, the discipline has produced important work on a wide range of topics and has dramatically reconfigured the philosophical understanding of issues such as pictorial representation, fiction and the imagination, the relationship between art and emotion, artistic ontology, the value of art, and the aesthetic appreciation of everyday practices and the natural environment. This body of theory and philosophical practice easily lends itself to the current issue. VR is a genuinely fascinating media development, and one that is of obvious interest to philosophers generally, and philosophers of the arts specifically. But while the academic and scientific literature on VR is large and ever growing, and there is also a reasonable amount of philosophy on virtual reality, there has been very little attention paid to the medium from within the field of philosophical aesthetics. Perhaps this lack of attention is partly explained by the very newness of VR; but also, I think, because aestheticians are not yet convinced about the relevance of this technological development to their typical concerns. I hope to change this.

Because this is principally an investigation borne out of the philosophy of the arts, the theoretical orientation of this book and the literature that I draw on to explore the topic will be somewhat different to most other accounts of virtual reality. Virtual reality will here be considered as a medium to be understood alongside of painting, film, photography, as a means of configuring representations of real and imaginary spaces, as conveying a sense of spatial experience itself, and as allowing for distinctive aesthetic and artistic practices. Thus, I will situate VR in terms of several key issues in the philosophy of the arts that relate to such media. I will be asking about the representational or depictive nature of VR; the historical and conceptual relationship of VR media to paintings, photographs, and other forms of picturing; the apparent realism of VR media and how it might relate to other "illusionistic" works such as *trompe l'oeil* paintings; the striking affective experiences that VR media conveys or evokes; and how the interactivity inherent in VR may reconfigure our experience of art.

It is worth introducing just a few of these issues at the outset of this study to give the reader a sense of how the consideration of VR media might affect long-standing philosophical issues. Pictorial realism was a significant issue in aesthetics in the 20th Century, often centering around whether the technique of linear perspective, developed in Florence in the 15th Century, really counted as more realistic than alternative ways of depicting space. Much has been written either defending pictorial realism or subjecting it to criticism. Adopting a favorable stance, Ernst Gombrich (1960) contended that linear perspective comprised an incremental refinement of depictive naturalism by employing the process of "schemata and correction" to approximate natural experience and perception. But there was also a lot of doubt about the naturalism of the technique, with Nelson Goodman contending that for a picture utilizing the technique of linear perspective to be seen as reproducing the geometry of the scene depicted, "the picture must be viewed through a peephole, face on, from a certain distance, with one eye closed and the other motionless" (Goodman, 1976: 12). For Goodman linear perspective was an artificial and moreover conventional means of spatial depiction, and he concluded that "the behavior of light sanctions neither our usual nor any other way of rendering space" (1976: 19). But virtual reality media loosen Goodman's constraints on the viewer, who may now, with both eyes open, move freely towards and around pictured objects in a way that, at first blush, gives a great impression of our natural experience of a visual scene. In this way VR seem to revitalize the potential for pictorial realism.

VR may also alter our emotional connection to the arts, and our capacity to be emotionally affected by what we find depicted within them. In traditional artistic media such as painting or film, the apparent objects, people, and events of our appreciation seem separated from us, existing in another space that almost never registers our presence as a viewer or allows us to influence what occurs within that space. But VR media give precisely the appearance that its users occupy the same space as the objects or events they perceive. The persisting issue of the reality or fictionality of the emotions we have for fictions—for example, feeling for Anna Karenina's plight or fearing a green slime slithering toward you "through" a cinema screen (Radford, 1975; Walton, 1978)—should surely be reconsidered given the development of a medium in which you have the seeming potential to occupy the same space as the objects of these emotions. A green slime might be worrisome on the cinema screen, but what if it is in the same room, slithering not just towards the screen, but towards you? I would imagine it could be utterly terrifying!

And finally, the ideas of *interaction* and *interactivity* have been the topic of a great deal of philosophical interest in recent times, largely because of the development of interactive media such as computer art and videogames (Gaut, 2010; Lopes, 2010; Tavinor, 2009). Works in these media are perhaps interactive in that the user is not merely an audience member, but also a kind of performer, playing a role and making decisions about how the scenarios they view play out. And again, VR media seem to frame this issue in an especially vivid way. In VR games, not only will the game be affected by your decisions and actions such that you might be described as taking a role in the gameworld, or interacting with characters, as in previous non-VR video-games, but you can *seemingly* physically gesture, reach out, and interact with the objects and characters in virtual reality. That is, the decisions you make as a performer may affect the world through the conduit of your *virtual body*. Having been terrified by the green slime, you might steel yourself and swing your virtual sword at the gelatinous beast.

In addition to these specific issues, virtual reality media connect to aesthetics at a more fundamental level which predates the current discipline of philosophical aesthetics, and even its origins in the Enlightenment. Before it was initiated as a discipline by the German philosopher Alexander Baumgarten and other enlightenment thinkers, *aesthetics* was principally a term deriving from the Ancient Greek *aísthēsis*, referring to sensation and perception. In this sense, the aesthetics of virtual reality media, rather than merely comprising the concerns of disciplinary aesthetics, will detail how VR media relate to sensation and perception on a more fundamental level. A significant part of this book will adopt this focus. The aesthetics of virtual media, in this sense, involves the study of how such media amount to a *virtualization* of our ordinary sensory and perceptual capacities, giving rise to our apparent perceptual agency in a virtual world. Indeed, I suspect that the aspect of VR of most interest to aestheticians will be its remediation of our experience of the external world, and the implications this has for the kind of medium VR is understood to be.

These are just a few of the fascinating philosophical issues provoked by the consideration of VR media, and there are many others as we will discover. I will return to these issues in later chapters, but I hope their introduction here should be enough to pique the interest of my readers. Of course, many issues in aesthetics, including those introduced above, are deeply contested by philosophers; I cannot solve these issues here, rather my hope is to show how VR can be profitably situated within these debates, perhaps influencing their future course. And I do intend to contribute to the understanding of these long-standing issues in this new setting because to my mind, the features found in VR media often align with, and justify, some positions on aesthetic debates over others.

Unfortunately, these issues have not much concerned philosophers when they have directed their attention to virtual reality. Instead, the most frequent theoretical orientation adopted by philosophers is on metaphysical questions such as the *reality* or otherwise of the apparent objects and worlds depicted by VR media. And in this, the discussion of actual VR technology and media typically only plays a small role. Rather, this orientation focuses on speculated cases of *perfect virtual realities*; that is, hypothetical virtual worlds that are indistinguishable from our own actual world. Current VR media are seen only as part of the *inevitable* technological drive toward such perfect virtual worlds. In adopting this orientation, VR media, rather than being the focus of the investigation, often becomes a prop for pursuing traditional philosophical concerns such as those associated with Cartesian skepticism about the reality of the external world. VR media thus becomes another means of exploring metaphysical ground previously traveled via the thought experiments of evil demons (Descartes, 1641), brains in vats (Harman, 1973), or experience machines (Nozick, 1974). Inevitably we encounter the question of whether we can tell if our current world *is not already a virtual one* (Bostrom, 2003).

There are, no doubt, interesting issues regarding the metaphysics or ethics of perfect virtual realities. Moreover, this might be an appealing way of conveying some of the traditional issues in philosophy to today's cohort of philosophy undergrads. These kinds of speculations about virtual worlds and objects have been problematic for the philosophical understanding of the medium of VR itself, however. First, they have effectively diverted attention away from a concern with the many fascinating aesthetic questions raised by virtual media. Metaphysics has the habit of dominating philosophical discussions of VR, and I have seen more than one promising discussion of VR aesthetics derailed by ontological questions about the reality of virtual worlds. Second, as we will discover later in this book, this orientation produces an enormous amount of conceptual complication and confusion, most of which strikes me as being unnecessary and avoidable.

One of my principal hopes here is to provide a simple account of VR media that avoids this metaphysical confusion. So apart from the second chapter, where in an analysis of virtual media I stake out my quite minimal ontological commitments, and the final chapter, where I will attempt to deflate the elaborate metaphysical claims of other philosophers, questions of metaphysics will not concern me here. Indeed, I will try to minimize reference to *virtual reality* itself, and one way I will do so is by employing the initialism VR, so that the descriptive content of the full term, and its apparent metaphysical implications, do not intervene on the discussion where they are unwanted or unneeded. In many ways it would be better to avoid the terms *virtual reality* and *virtual worlds* altogether, though because of their terminological predominance, that strategy would be unlikely to succeed.

There are additional pitfalls here for the philosopher of the arts. Foremost, because of the media and theoretical buzz around the topic of virtual reality, and because as an experiential phenomenon they are indeed quite striking, we are likely to think virtual media to be unprecedented or to be more of a radical departure from previous media forms than they actually are. Thus, I will take care to identify what is genuinely novel and surprising about VR media, while at the same time recognizing that these media are a part of an historical development, and indeed that the features that inspire some thinkers to make quite radical claims about virtual reality are precedented in much older forms of representational media.

1.3 The Book Ahead

This book involves an analysis of virtual media, how they relate to previous representational forms, their potential and promise as a depictive and artistic medium, and how they might reconfigure or virtualize our experience of the world. I thus need to spend considerable initial effort characterizing VR media to provide a clear target for this discussion. Chapter Two begins this task by enquiring into the concepts of a *medium* and of *virtuality*. Focusing my analysis on current practice and technology, I will first explore the various senses in which some object, activity or experience might involve a medium, finding there to be two key related senses that seem almost ontologically primitive to our understanding of the world. Virtuality is itself a contested term; I will examine competing views, ranging from interpretations that hold *virtual* as most informatively contrasted with *real*, and others that see *actual* as the relevant conceptual contrast, eventually settling on the latter interpretation. It will turn out that the idea of *remediation* is crucial to understanding virtual objects, activities, and experiences, and so the concepts of virtuality and media, initially separated in this chapter, must ultimately be considered in conjunction with one another. In practice it is the instantiation of the functions or structures of an object, activity, or experience within an unfamiliar or novel medium that comprises the basis of *virtualization*. A virtual store, to take a very basic example, is one where the functions of a brick-and-mortar store—browsing items, placing them in a cart, checking them out and paying for them—are transferred with minimal loss of function into a new digital medium. Several more examples will be inspected to illustrate and justify this claim.

If this is what it is to be a virtual medium, how do the items that are the focus of this book, particularly VR headsets, count as such? Which functions or structures do they preserve through media transposition? Chapter Three takes the account of virtualization developed in the previous chapter and applies it to VR media. This will be a largely technical chapter that explains how VR media allow for a combination of virtual perspective and interactivity, amounting to the virtualization of natural spatial experience. But given the conceptual refinement of the previous chapter, where virtuality was defined in terms of the media transposition of structure and function, and not tied to any specific medium, the story starts earlier: the discussion will begin with an apparent digression on linear perspective in painting. This diversion will allow us to inspect a key element of the evolution of virtual perspective, though one that employs a non-computer medium. From here the chapter charts additional elements of virtual media, including 3D graphical picture spaces and the virtual camera, motion tracked perspective and stereopsis, and non-visual media such as egocentric sound spaces, and haptic and kinesthetic means of depicting space and perspective.

I will thus claim an historical and etiological link between visual VR media and the development of linear perspective in painting. However, this raises the question of whether VR visual media do comprise pictures. Considered in terms of previous theories of pictorial seeing, there are several reasons to think they are not. These issues derive from the popular notion that picture perception involves a kind of "twofold seeing": that is, somehow seeing the surface of an image, and seeing through this surface to the items depicted. Aspects of VR visual media may be at odds with this theory, including that they appear to present the user not with a simple surface, but with an *encompassing visual field*; their potential for *egocentric* seeing in which the user perceives their own apparent spatial relationship to the items pictured; and finally, the user's apparent ability to interact with items in the visual field. Chapter Four argues that VR visual media can be reconciled with twofold theories of picturing, but that those theories themselves might not come out unscathed. The chapter ends by addressing the interactivity of VR media, framing the beginnings of an answer to what this interactivity amounts to. The 3D visual configurations seen in the surface of VR pictures are associated with interactive features in a way that makes the medium an interactive one, but also gives the impression that the user interacts with virtual objects or worlds.

Chapter Five addresses the typical and potential uses of the interactive egocentric pictures thus produced by VR media. VR is perhaps most commonly associated with videogaming, but this does not exhaust the uses to which it is put. As well as the VR fictions typical of videogames, VR can be used to document the real environment, or even allow the seeing and interaction with real objects that are present to the user. The egocentric picturing of a VR headset may thus place the user within the fiction of a mysterious and darkened house, allow the inspection of a current or historical site such as New York City or Ancient Rome, or even permit the user to see and catch a real ball that is tossed to them. This variation in uses not only upsets some of the assumptions that are frequently made about VR—for example, that the objects depicted by VR are *not real*—but it also demands that we expand on our theoretical resources and terminology. We must look at how VR plays a role in depicting fictional scenarios and how the documentary uses of VR media involve the remediation of our perceptual and interactive engagement with the actual world. This section will thus initiate the theses of *virtual fictionalism* and *virtual documentary*, largely characterized by a difference in their intentional objects. The chapter then details some especially surprising cases of virtual documentary: we will see how VR media also allow users to really see and interact with objects, even those from which they are physically dislocated, thus constituting a kind of prosthetic seeing. I will refer to these as instances of *virtual transparency*. Finally, this chapter considers the artistic uses to which VR media are currently being put, and their future artistic potential.

As noted at the outset of this book, one striking aspect of VR media is their apparent realism. Chapter Six tackles VR realism head on. There is an initial plausibility to the claim that VR media are more realistic than previous representational forms, and I will muster further anecdotes and some empirical evidence to this effect. However, we will also discover that these claims of VR realism are ambiguous. After an initial discussion of how the concept of realism has been treated in the aesthetic tradition, I will cover several possible explanations of the apparent realism of VR pictures, arguing that some are more useful than others in justifying the initial plausibility of VR realism. That VR media are distinctly *immersive*, and that this constitutes their realism, will be subjected to critical scrutiny, a scrutiny from which the popular concept will not emerge intact. I then look into what might be called *perceptual or psychological realism*: the idea that virtual media engage our perceptual and affective psychology so as to seem real to us. This category includes the familiar phenomenon of *spatial presence*, but also a wide variety of other responses. Finally, I attend to the problems or challenges to the idea that VR media provide an especially realistic rendering of spatial experience. In practice, the illusion of VR is severely compromised by technological shortcomings. But there are also deeper conceptual questions about the possibility of realism in VR media. Revisiting the debate about the apparent realism of linear perspective, I raise the skeptical possibility that VR picturing media constitute a largely conventional and hence non-realistic mode of spatial picturing, in much the same way as linear perspective might be considered conventional.

Chapter Seven finds me finally addressing metaphysical realism about virtual worlds and objects, a topic I have tried to fend off throughout the preceding chapters, and often without argument. But having now given a theoretical account of VR as comprising an egocentric interactive mode of picturing, I will attempt to *deflate* the metaphysical issues that dog the analysis of virtual reality itself. The chapter covers several forms of what I will refer to as *ontological realism*, that is, the idea that virtual objects and virtual worlds are real things in some sense. Firstly, I argue against the common claim that there is an *ontological spectrum or shift* from the perception of the real world to the perception of virtual worlds, and that this shift involves something like the adoption of beliefs in the existence of the latter. Next, I will address the claims that apparent virtual objects have a real existence either as computational or digital objects, or as socially constructed objects of the kind that are common in philosophical accounts of social kinds. Finally, in this book, noting the prevalence of such realism about virtual objects, I try to provide some reason, other than the truth of such claims, for why they might be so common. My *error theory* for virtual onto-logical realism will be partly terminological, in that I will claim that the very locution of "virtual reality" has had an unfortunate influence on the philosophy of VR. But more substantively, I will argue that if one does not take care to distinguish between the medium of VR picturing, and the range of intentional objects arising from its varied use, then one can easily become confused about ontological matters.

Ultimately, there is nothing especially metaphysically provocative about the egocentric and interactive functions of VR media. In fact, if I were to frame the thesis of this book in a very simple way, it would be that "VR is a technologically fancy kind of picturing." To see this, however, we really need the full theory of VR media in front of us, because it is all, admittedly, a tiny bit complicated. Let us get started then!

2 What Is a Virtual Reality Medium?

2.1 Towards an Analysis of Virtual Reality Media

My principal intention here is to investigate VR as a medium—or more likely, as a collection of media that are frequently bundled together into a suite—that allows for the representation or documentation of real and imaginary worlds, and the wide range of interpretative, interactive, and practical implementations allowed by such representational models. Plausibly, virtual reality media are a new kind of thing, allow a distinctive style of experience to their users, and give a profound sense of experiential realism. However, to evaluate the claim that VR media are a new, distinctive, and realistic medium, and provide a theoretical account that might substantiate these claims, we need to understand what they fundamentally are. Given the wide and potentially inconsistent usage of the term "virtual," some sharpening of concepts is needed here to see why the virtual reality media picked out for attention above are appropriately grouped together.

In the introductory chapter, I identified the topic of this study by pointing to a range of items that I found interesting and worthy of study, particularly stereoscopic headsets, virtual audio, and spatial tracking. These I referred to as *VR media*. Philosophers call this kind of "definition by pointing," the practice of ostension. *You see these things here; these are what we are concerned with; these are VR media.* Unfortunately, there are problems with ostension if we consider it as a reliable means of defining or even identifying clear and unequivocal categories of things. *Just why are you picking out these things, and not others, as virtual reality media?* There are, after all, a diverse range of devices and peripherals besides stereoscopic headsets, that are associated with virtual reality. Omnidirectional treadmills that allow users to turn and walk in a virtual environment and tracked gloves that give tactile feedback are just two examples. Furthermore, many such devices appear quite like devices we would not necessarily point to as cases of virtual technologies: the old-style *View-Master* picture viewer allows for stereoscopic viewing of visual scenes but does not seem like a fully fledged virtual technology.[1] Nintendo's notorious *Powerglove* also has a superficial resemblance to virtual reality gloves, but lacks their VR functionality—in fact, it lacks almost any

DOI: 10.4324/9781003107644-2

functionality whatsoever! What is it that justifies the intuition that haptic gloves are VR media, but the Powerglove is not?

An obvious response, of course, is to first characterize *virtual reality* itself, and then to identify virtual reality media as those media employed for the purposes of depicting virtual reality. But here the problems are also significant, because so many things have been labeled as cases of virtual reality, virtual worlds, or the like, that we might simply become confused about which *really* count as instances of virtual reality. Early virtual reality theorist Michael Heim identifies this issue when he notes that the "many offshoots" of virtual reality might distract us from the "central meaning" of the term (Heim, 1998: 4).

> Many contemporary experiences—from using ATMs (automated teller machines) to visiting Disney's "Star Tours"—serve up, in a variety of ways, the experience of interacting with simulations. What we call the automated teller machine is not truly a bank teller but a machine that performs many of the functions of a bank teller. The "as if" quality—following the dictionary definition of "virtual"—qualifies the ATM as a virtual bank teller. So pervasive are these simulations that we find "virtual" describing everything from phone sex to the kind of non-committed gaze used while walking past store windows in the shopping mall. These extended meanings are all interesting, and even important, but we should not let the extended meanings confuse us.
>
> (Heim, 1998: 4–5)

But even Heim's putative "central meaning" of the term might be elusive. Having surveyed the diverse recent literature on VR it would be fair to say that there is no great agreement about what virtual reality is, how it should be defined, and whether we should even refer to it as virtual "reality" or by some other name.

Indeed, there is a terminological looseness across the discussion of VR, including much of the philosophical work to be introduced here. There is, for example, the question of whether the terms "virtual world" and "virtual reality" are equivalent and interchangeable. In the philosophy of technology, and in games studies, *Second Life* is by now the most tired example of a *virtual world* because of how frequently it has been cited as an example of such since its launch in 2003. *Second Life* is displayed on a 2D computer screen like most videogames, and it involves features like a persistent 3D world model, avatars that one may not only control but also adopt as identities, and the allowance for users to interact with each other. But unfortunately, this virtual world has also often been introduced in the context of *virtual reality* without ever acknowledging the media particularities of the latter (see for example, Ludlow, 2017). Credibly, virtual reality involves something more, that is, the feeling of "being in" such a place. Other than by using third-party developed VR hardware and apps, *Second Life* does not involve VR media.

There have been recent calls for increased attention to be paid to the definition of the term "virtual reality" to allow for the discrimination of these persistent ambiguities (Kardong-Edgren, et al. 2019). There are some rudimentary definitions available. For example, the theorist Michael Heim has had a large influence over the theory of VR, and his definition of virtual reality as "an immersive, interactive system based on computable information" is one instance of this influence (1998: 6). The definition is revisited in recent work on virtual reality by David Chalmers in his claim that "a virtual reality environment is an immersive, interactive, computer-generated environment" (2017: 312). Heim's definition, and Chalmers' version of it, does a reasonable job of characterizing the prototypical cases of VR, but as an analysis of VR, it is wanting in several regards. First, the terms of the analysis themselves need explicating. What is it to be *immersive, interactive,* and *computable*? These concepts themselves are likely to be slippery, as we will find later in this book. Second, do these qualities, explicated in a reasonable way, or revised, really characterize all cases of VR? I am not going to follow this lead and attempt to give a formal definition of VR, because I do not have any great hope that this could be achieved, and I am also not sure that the pay-off would be worth the effort in the context of this book.

Nevertheless, the identification of the concepts of immersion, interaction, and computation—or something like them—has a heuristic value for my study. Chalmers also does us the service of identifying "virtual reality technology [as the] technology that sustains virtual reality environments" (2017: 3). I take this to broadly encompass much of what I have referred to here as virtual reality media. And so, an alternative way of framing my concern in this book is the question of what these *immersive, interactive,* and *computable VR technologies* comprise. These additional requirements may make it more obvious about exactly where in the world we should be pointing to identify the object of our study, and they also provide more theoretical substance and an extra set of questions to answer. What is it about immersion, interactivity and computability, and their combination, that generates the phenomena identified above as VR media? How well do these terms stand up to critical scrutiny and how might they need revision? Can these concepts be discerned in media that pre-date VR, and if so, how might this influence our understanding of VR itself? With this refinement of the issues and acknowledging and setting aside the urge to attempt to define VR, I will follow the lead of the ostensive act of picking out the obvious cases of such virtual reality media and technology, and immediately theorize about these things.

But how to theorize such virtual media? The literature on virtual reality and its characteristic media and technology is multi and inter-disciplinary, and theorists have grounded their work in any number of bodies of method and theory, some quite elaborate, and still others obscure. The method in my study, however, is fundamentally analytic. We will be prudent to pay heed to a familiar philosophical imperative, "to divide each of the difficulties under examination into as many parts as possible, and as might be necessary for its adequate solution" (Descartes, *Discourse on Method*, Part II). When faced by complicated and potentially confusing issues, we should break them down into their simpler component parts and see if in this analysis, the broader issues become clear and can be reconstituted in an illuminating way. The relevant components I have chosen for the following analysis are the concepts of *virtuality* and *media* (putting aside until much later the pesky term "reality"). Let us approach these fundamental terms in reverse order.

2.2 What Is a medium?

The concept of a medium is quite general and in many respects is foundational to our understanding of the structure of the world and our agency within it. As a noun, "medium" has several related and frequently intermixed senses, but the two basic senses of interest here are that of a medium as first, *an intervening substance through which things traverse or are conveyed, and second, a physical means of performing some action or achieving some end.* The word itself derives from the Latin, *medius,* meaning "middle" or "moderate." The development of this Latin term into its modern wide usage— particularly to take on the meaning of an intervening substance—seems to have coincided with the development of modern science, in which the concept of a physical medium played a frequent and productive role in physics. Its cognate in Ancient Greek is μέσος, which in its modern Greek form μέσο, may signify "middle," but also "means," "medium," "tool," and even "agent."

In a first sense, the concept of a medium is a physical notion concerning how forces, properties, and qualities can be instantiated in, and be transmitted through, different kinds of matter. Matter comprises the fundamental constituents of the universe, the stuff of which things are made. A medium is not just a kind of constitutive matter, but a constitutive *middle ground* through which a phenomenon might pass or be transmitted. As such, it allows for a physical event or thing to retain its identity even as it physically traverses a medium, potentially traveling into another. A good example of this sense of media involves the notion of a *physical force* moving through matter. Longitudinal and transverse waves of kinetic energy, generated by the slip or rupture of a fault in the earth's crust, and propagating through the earth as a violent shaking, constitutes an earthquake. At the surface of the earth these waves of energy may erupt into the air as a bellowing roar. The nature of the medium thus has a structural and phenomenal impact on the nature of the waves of kinetic energy: in the earth they comprise a violent shaking, capable of bringing down buildings and injuring or killing people; in the air, because the longitudinal compression wave may move more quickly through the medium of the earth than the transverse wave that is responsible for the most violent shaking, the buzzing bellowing sound produced by the longitudinal "P-wave" may herald the oncoming destruction. At any one moment the earthquake exists in a particular kind of constitutive matter, but it may move beyond this into other media.

In a second sense, this notion of a constitutive *middle ground* operates in an agential or intentional context. Thus, in addition to referring to a transitional material substrate, a medium may refer to a physical means of achieving an aim, or more simply, a way of doing something. This intentional interpretation of a *medium of action* thus gives the impression of being an intentionalist corollary of the physicalist sense: we understand that a physical event might have different material instantiations and may transition from one media embodiment to another without the loss of its identity, and it is natural to see that intentions or goals too may be satisfied in many ways. As the saying has it, there is more than one way to skin a cat. The concept thus allows for the instantiation of *functional* artefacts in different materials: a knife, an instrument for cutting, can be injection molded in plastic, pressed out of stainless steel, cast in bronze, or even shaped out of strips of wood. These differing material instantiations need not affect the identity of the items *as knives* but do affect their qualities, and hence their aptness for a given purpose: a wooden knife is fine for spreading butter, but not so great for carving a turkey. There are, of course, materials that are unlikely media for the production of a knife: a knife made in jello is unlikely to function as a knife at all, even to the extent that we might question the status of the object as a knife, perhaps thinking it a mere imitation of one.

Medium then, is an ontological pattern that bridges the material and intentional worlds, that allows us to conceptualize these overlapping domains. In both domains, x's may be instantiated or satisfied in different media, and x's may retain their identities through such media traversal. A good general example of how these senses overlap is the concept of a storage medium. Foremost, storage media are a means of achieving the goal of information storage and retrieval. A physical storage medium is a means of encoding information or data, often in the form of a physical device such as CD-ROMs, flash drives, and hard drives. In data storage, information, whether constituting a computer program, an image, or a text document, is compiled into a form that can safely be stored and later retrieved in its original configuration without loss (or with minimal loss) of its original structure or function. Information in one storage medium can also be converted to be stored in another. What is instantiated in this case is the information itself, the medium being one of many physical means of embodying information. Being reducible to the values of defined parameters, information is particularly amenable to being instantiated in various media because there are any number of ways to physically embody such states. In analogue forms, we find magnetic tape where information is stored in the variations of the magnetization of an iron oxide coating on a plastic tape. In a digital form, such as a compact disc, the information is encoded in optically registered indentations on a polycarbonate disc. Computer memory often involves the storage of information in the form of semiconductor memory where the parameter values encoding the information comprise the electrical values of gates in a complex circuit. Physically, such memory involves the controlled oxidation of a semiconductor, usually a silicon chip. In all such cases of information storage, a transduction mechanism is required to store and retrieve the information in the material substrate. The advantage of computer chip memory is that the storage and retrieval of memory is first, very fast, and second, *random*, in the sense that unlike a compact disc where information is retrieved consecutively and in the same sequence in which it was written, in chip memory any discrete part of the information may be retrieved in any sequence.

Several other specific cases of media are relevant to understanding VR. First, we can note that the concept operates quite naturally in the mental realm. A *mental* or *sensory medium* may be conceived as the substrate of our impressions or ideas about the world, either as we receive these impressions from the external world through sensation, or as we ourselves formulate our ideas or thoughts within our minds. This sense seems much like a mentalistic correlate of the physical sense of a medium as an intervening substance through which something—in this case impressions or thoughts—are conveyed. And indeed, a link between physical and mental media has often been quite close in philosophy, for example in the mechanical theory of sensation formulated by Descartes. Descartes conceived of the visual sensation of qualities such as colors, as involving corpuscles of light, traveling from luminous objects and producing through their action, ideas "formed in our imagination through the inter mediary of our eyes" (Descartes, *The World*, I). The ideas and thoughts produced by these sensations might themselves be conceived of as existing in a medium, perhaps linguistic (Fodor, 1975).[2] Conceiving of sensations, experience, ideas, and thought then, all quite naturally partake in the concept of a medium, and that concept might naturally shape how we subsequently characterize and understand the subject matter.

And, of course, our thoughts can be communicated outwards, principally through speech, gesture, and writing, but also through many other expressive forms. Thus, again drawing on the physical sense of the concept, communication can be conceptualized as the dissemination of ideas through a medium. This now utterly familiar view of communication leads us to think that words from our mouths might travel on the air, that we might set our ideas down on a page, or that gossip may travel through the grapevine. An early use of the concept in this sense can be found where Francis Bacon discusses "the expressing or transferring [of] our knowledge to others":

> For the organ of tradition, it is either Speech or Writing: for Aristotle saith well, "Words are the images of cogitations, and letters are the images of words"; but yet it is not of necessity that cogitations be expressed in the medium of words. For whatsoever is capable of sufficient differences, and those perceptible by the sense, is in nature competent to express cogitations.
>
> (Bacon, *The Advancement of Learning*, XVI, 2)

The final sentence, particularly in its reference to "sufficient differences," I take to mean that any perceptible item capable of exhibiting kinds and degrees of variation, may act as a medium of thought. So just as the material instantiation of a knife requires specific features of a physical substrate—namely, a material that can "hold an edge"—the communication of ideas makes formal requirements of its material substrate.

With the advent of technologies such as the printing press, communication media took on a prominent role in cultural development, eventually leading *The Media* in the form of print, radio, and television broadcasting. These comprise information channels deriving from specific technological advancements, and they have recently been joined by digital media, an event that has radically reconfigured our ways of gathering and distributing information. For the reason that their displays comprise text, audio, images, video and animation, digital media can also be referred to as "multimedia" (though this term already has a bit of an old-fashioned sound to it). The effects of digital media on the prevailing modes of personal communication in the form of email, online personal messages, video calling and meetings, and social media applications, have been perhaps the most significant aspect of the virtualization of our modern world. However, while digital media are an important part of the background to this current study in that VR media may be among their number, they cannot take much of the focus here.

Digital media and VR media ultimately owe their existence to the more fundamental concept of computational media, a category of media that clearly does require attention. Abstractly, a computational medium is one in which tasks are achieved by the performance of sequences of logical or arithmetical operations, often composing programs comprising sets of more basic functions. The tasks that can be achieved by such programs, as we know from the advances in technology that are reshaping the world around us, are breath-taking in their scope: computers keep our planes in the air, allow us to access a vast amount of human knowledge on our smart phones, and remind us to go for a walk if we sit on the couch too long. Most immediately, the keyboard on which I write these words, and perhaps the screen on which you read them, are surfaces through which we interact with a computational substrate. And, in VR, computational media may give users the impression that they perceive and interact with alternative worlds displayed on screens before them. But all these cases depend on the very basic idea of *computability*: a task or problem is computable if its solution can be arrived at by the performance of a series of discrete logical or arithmetical operations, or what is called an algorithm.[3] Computer programs are built of such algorithms, and when given a material interpretation—that is, where their inputs and outputs are mapped to interfaces or peripherals that are integrated with the physical world—they are a medium that allows for the performance of physical tasks such as flying aircraft and reminding me to go for a walk. Computers are thus *imitation machines* in that they may imitate functions ordinarily found in other media, devices, or indeed, tasks usually performed by people. This imitative potential of computational media is going to be central in understanding VR as a medium.

Finally, of relevance is the particularly rich intermixing of the concept of a medium as it functions in the description of the practices of art. The concept is employed in Aristotle's account of imitation in the *Poetics* where the media of drama—rhythm, language and harmony, or their various combinations—allow it to imitate life:

> For as there are persons who, by conscious art or mere habit, imitate and represent various objects through the medium of colour and form, or again by the voice; so in the arts above mentioned, taken as a whole, the imitation is produced by rhythm, language, or "harmony," either singly or combined.
>
> *(Poetics, Part I)*

Here we already see the idea of an *artistic medium*, which we now understand as comprising not only rhythm, harmony, and language, but many things besides, including technologically sophisticated media such as photographic images, recorded sound, and interactive computer animations.

Artistic media can be conceived as the practical or physical means of achieving artistic intentions. Most literally, or perhaps practically, we might consider the media of art as the physical means through which artworks are made, and the things that compose the physical structures of artworks. The artist's paints are her medium, as is the pictorial medium she employs in a more abstract way, and the pipe she thus paints—though it is not a pipe—is depicted via pigments laid down on a canvass with a brush. A matte medium—a substance used by painters to increase the flow and translucency of paints or achieve a matt finish—embodies its artistic ontology and function in its very name. In music, a drummer's medium are the pulses of sound she collects and distributes into rhythms, but also the physical instruments she uses to produce these sonic events: tom toms, cymbals, snares, and hi-hats. And, of course, these physical things are a means of her artistic intention to set down a steady groove for the band she plays in. Through the means of orchestral instruments and conventions, a symphony can *express* despair and triumph, perhaps within the same movement. The embodiment of these expressive features in the medium can be subtle or even mysterious, entirely evident to the listener, but puzzling to the theorist. And art can also be considered to achieve representational aims and can portray many things through its varied physical media: the Napoleonic wars, a day in the life of an Irish salesman of newspaper advertisements, the story of a young man leaving his home on the planet of Tatooine and discovering his true identity. Some artworks, too, may traverse different media instantiations as when a musical work is transcribed for performance on alternative instruments.

It is in artistic media that we come closest to the sense of medium to be used in the following theory, particularly in my focus on the predominating visual nature of VR as a *pictorial medium*. Both the physical and intentional components inherent in the idea of artistic media will be crucial to my account of the aesthetics of virtuality, though much of the focus will be on how these have been transformed by computational media. I will explore many of the physical and technological aspects of virtual media, including pixels, lenses, virtual cameras, but also the purposes and intentions that these virtual media are designed to fulfil, whether they be the interaction with, or documentation of, the real world, or the construction of props for our imagination.

2.3 What Is Virtuality?

What does the concept of virtuality itself mean? That is, how does the word "virtual" modify concepts such as "reality," "world," "environment," and "media" when it is applied to them? As noted earlier, the wide and diverse use of "virtual," particularly in the time since the computer revolution, tends to obscure the meaning of the term. The digital revolution has given rise to virtual meetings, virtual stores, virtual currencies, and any number of other virtualizations of items, functions, or activities that we previously and customarily encountered in the familiar world beyond the screen. So, like the concept of *medium*, in the analysis of *virtual* we begin with a profusion of uses and a conceptual vagueness that may simply confuse us. Given the extraordinary number of things that have been labelled as *virtual* in recent times, and that such references are often extraordinarily vague, one might have doubts about the real utility of the term, perhaps even suspecting it of being a fashionable though vacuous way of referring to our current technological predicament. So how might we best get a grasp on the central meaning of the term in the formulations of virtual reality, virtual worlds and environments, and virtual media?

The obvious lead to follow is to begin with the dictionary use of the term noted by Heim. The common sense of the term "virtual" might be most succinctly framed as "as if but not strictly so," in the sense that a virtual x might for practical purposes be treated *as if it were an x* even though it differs in certain respects from a real x, often in virtue of lacking some property or condition that a real x normally has. Or, in an alternative formulation owing to the Oxford dictionary, virtual can mean "almost or nearly as described, but not completely or according to a strict definition." For example, when Chomsky notes of the apparent doctrinal orthodoxy of writers and opinion makers in the news media, that "there is kind of a fundamental conformity, which is a "virtual" requirement to enter into the media," he acknowledges that there is no real such contract to conform, but that for practical purposes there might as well be (French, 2010). The sense of the word is clear enough in this instance and others, but how this sense applies in the case of virtual media is precisely what needs to be settled here. In the current section I am going to formulate my answer to this question. But first, let us clear away a couple of tempting missteps.

A first potential error in the analysis of the core sense of "virtual" is the temptation to think the term is more or less interchangeable with "computerized," so that a virtual medium is merely a computer medium. Here, a virtual item or event would be "almost or nearly as described," but not completely so in virtue of utilizing computer technology. And in this sense, the word "virtual" admits additional synonyms such as "digital," "online," "internet," "networked," or even, in an increasingly antiquated way, "electronic." This usage is sometimes listed in dictionaries as a secondary sense of the word "virtual,"as in the Oxford definition of computational virtuality as "not physically existing as such but made by software to appear to do so." In this way, *virtual learning* is learning that employs computers and online networks for teaching, the distribution of learning materials, and the assessment of students, and not a classroom setting which was the customary location for such activities; a *virtual meeting* is one that occurs not in person but online via an app such as Zoom or Skype; a virtual forum is a message board and not a physical civic space. Hence, a virtual medium is simply a computer medium. And of course, the profusion of the term "virtual" during the COVID-19 pandemic can be attributed to the radically expanded use of such online formats, and the terminological fashion to refer to these as virtual technologies.

But while much of the popular use of the term does seem to be of this form, and the rise of computer media has certainly expanded the prevalence and awareness of virtuality as a concept, it would be a mistake to identify virtuality so closely with computation. Collapsing the two concepts would obscure an important distinction. If we reach back to the core sense of virtual as meaning "as if but not strictly so," or that virtual *x*'s might be practically treated as *x*'s, while lacking features of genuine *x*'s, then not all computer media need count as virtual media. Computer applications such as internet browsers are not "practically treated" as some other kind of precedent artefact, as though there were *non-computer internet browsers* predating them. Rather, the internet is itself a computational phenomenon that generated the need for entirely new functional artefacts such as browsers. Internet browsers might be *compared* with earlier items such as microfiche viewers or slide shows, or in some respects might seem *similar* to them, but they are not in any sense functional correlates of these but achieve something entirely new. It is credible that a great deal of computing technology is like this in having simple functions, and we should not be tempted to refer to these as virtual at all.

More contentiously, perhaps, it may be that not all virtual media are computer media. If a virtual technological medium comprises one that instantiates an activity or process in an "as if but not strictly so" way, then many technological formats might count as such, including mechanical and analogue electronic formats. Zoom counts as a virtual technology in virtue of allowing for meetings at a distance, by capturing and conveying visual, audio, and textual representations of group interactions. People may thus meet "at a distance," while not being physically present to one another. But then, by the same token, a telephone call allows for *virtual conversation* in virtue of transmitting verbal communications between people who are distant from one another. The word "telephone," of course, as derived from the Greek, means "distant voice." Originally, telephones achieved this function by transducing the vibrations of the sound of a person's voice into an electrical signal that could be transmitted over long distances via wire, where they would again be transduced into the sound of a voice. Thus, telephones are a virtual technology, but not a computer one. Admittedly, identifying telephones as a virtual technology might sound a little non-intuitive, but this may merely reflect the familiarity of telephones themselves, and the frequent conceptual identification of *virtual* with *computer* that I am trying to undermine here.

If the reader is resistant to the idea that telephones are a virtual technology or medium, there are other examples of non-computational virtual media that might be more compelling. One such case is aircraft simulators. Most current aircraft simulators, such as the spectacular recent iteration of *Microsoft Flight Simulator* (2020), are near-canonical instances of virtual media in that they involve simulation, and they are run on computers. However, some of the earliest simulators were mechanical devices. The Link Trainer, produced from the 1930s to 1950s by Link Aviation Devices, was used to train pilots in instrument flying. Looking like small fairground aircraft rides, these trainers comprised reasonably complex and faithful representations of many of the flight systems needed to simulate instrument flying. It seems credible to me that these are cases of virtual flight simulators, and that they involve a kind of virtual "display" not entirely dissimilar to modern flight sims such as *Microsoft Flight Simulator*, even though they are not computational, or screen based. The "display" in the Link simulator is rather a physical, mechanical, and electronic simulation of an aircraft, featuring instruments, controls, and even a pneumatic tilting mechanism that gives the impression of the physical movements associated with flight control. Of particular interest is the gyroscopic attitude indicator or "artificial horizon," which visually displays facts about the virtual flight, and might itself count as a "virtual horizon" in both virtual and real aircraft. We can even imagine how, on similar mechanical principles to an artificial horizon, a staged pictorial backdrop outside of the cockpit windows could be developed to give a sense of a 3D visualization, depicting a kind of virtual world. Thus, we can already see here that the *computational* is not co-extensive with the *virtual*. Under the analysis that I will offer here, some computer media will be non-virtual; and some virtual media will be non-computational.

The second misstep in the analysis of "virtual" is placing too much emphasis on the distinction between the virtual and the real. Rather than relating it to computation, many theorists of virtuality, perhaps because they are influenced by the predominating use of the phrase "virtual reality," are tempted to characterize the virtual as primarily an ontological classification, and specifically, as one contrasting virtual things with real things. The meaning "as if but not strictly so," is thus interpreted as "as if, but not *really* so." In much of the philosophical literature on virtuality, the starting point or presumption is that *virtual* is a metaphysical concept and that virtual items have a different ontological standing than their real counterparts, perhaps being "less than real." So, for example, while he eventually develops a quite sophisticated position on the nature of virtual objects, the philosopher Philip Brey orientates his discussion by noting the "common belief that objects in virtual environments are not real but are mere imitations of or simulations of real objects" (2014: 42). Similarly, Pawel Grabarczyk and Marek Pokropski suggest that "relatively stable" to many definitions of the term "virtual," is something like a "non-physical representation of something physical," which "is only imaginary or conceptual" (2016: 26). They conclude that "it might be best to think about virtual as a kind of fundamental remediation of one ontological domain into a different ontological domain. 'Virtual' is thus very similar to the way we use 'mental' or 'imaginary'" (26). In these cases, it seems like virtuality is a mode of being, and potentially an imaginary, fictional, or merely conceptual one.

It is unfortunate that *virtual* is so frequently taken to be conceptually contrasted with *real*. I am going to leave my main argument against taking virtual reality to be a metaphysical phenomenon until the last chapter, because I want to avoid getting fixated on these issues here, and I want to have my alternative theory evident so that we will be able to see the failings in the metaphysical approach when we get there. But something must be said now about the dangers of analyzing the basic concept of virtuality in such metaphysical terms. The basic argument is that "virtual" should not be conceptually contrasted or defined in terms of being "unreal" "imaginary" or "conceptual," because many virtual items—virtual stores, virtual currency, and virtual memory—are simply real, and not unreal, fictional, imaginary, or merely conceptual. A virtual store such as Amazon is after all an effective way to purchase goods, and there is nothing fictional or unreal about this. So, to be virtual is not to be conceptually opposed to being real, and this is also the case with the virtual media ultimately being analyzed here: stereoscopic headsets, digital environmental models, pieces of code on a computer, are all very real things.[4]

Understandably, the starting point that virtual items are somehow unreal or merely imaginary quickly leads into metaphysics. This is particularly the case when the focus shifts from virtual technologies and media to the apparent items depicted by those technologies: that is, the apparent virtual worlds, environments, objects, and selves that loom so large in the philosophy of virtuality. Taking the virtual/real dichotomy as the relevant conceptual contrast, but then rejecting the idea that "virtual" equates to "imaginary," as I do here, some writers shift quickly from the credible claim that virtual media are real, to the very different claim that the items represented in such media are in some sense real. This leads Chalmers to make the startling claim that "virtual reality is a sort of genuine reality, and what goes on in virtual reality is truly real" (2017: 309). This, of course, is the point where I will need to suspend the discussion until it can be taken up in the final chapter, because we need the theory set out in the following pages to properly appreciate the fault in reasoning here.

So, having identified these potential missteps, how do we get onto a more secure footing? The answer, rather less obvious than it might initially seem, is that the correct conceptual contrast to draw against virtual is not *real*, but *actual*.[5] I believe that if we contrast virtual with actual, there is a reasonable way to refine the term to see its genuine theoretical utility in the case of virtual media, and indeed beyond this kind of usage. The basic definition of *virtual* introduced earlier is revised to *almost or nearly as described, but not actually so*. Now this demands the question, what is it for something to *be almost or nearly an x, but not actually so*? There are two key parts to my explanation: first, we must explain why a virtual *x* counts as a kind of *x*, and second, we need an account for what *actual* means in this formulation, and so what a *non-actual x* might amount to, and we must do so without equating actual with real, a conceptual option that has been rejected above.

For the purposes of this discussion, let us look at an example. Take an event like the Cold War and its proxy wars, particularly the Korean War. These events might usefully be considered a *virtual war* between the USA and the Soviets, and we might say that during these years, these combatants were *virtually at war*. Now, it would be odd to ask whether the Cold War was real: it certainly was, and a great many people died and suffered because of the war. It was not, however, an *actual war* between the USA and the Soviets, as war was never declared between these main combatants. This is the case even if on occasion they directly fought each other, as they did during the air war over Korea after April 1951 (Halliday, 1993). So, the question becomes not one of whether this was real, but why it was a war of any form, and secondly, why it did not escalate into an actual war. The answer to the first part of the question, is that it had the *efficiency* or *function* of a war between the USA and the Soviets: territory was contested, armies invaded and were pushed back, political and ideological will was asserted. But secondly, for very good reason, this was achieved in a *non-customary* or *unfamiliar* way in a proxy territory. In place of the actual armies of the virtual combatants were proxy forces, South Korea and ostensibly the United Nations on one side (though in reality, largely the United States) and North Korea (with the varying involvement of China and the Soviets as the war evolved) on the other. In this virtual war something was *retained* that warrants the reference to the event as a war between the superpowers (the geopolitical functions of war) and something was *lost* in virtue of which the war became non-actual (direct acknowledged combat, and the resulting risk of the war elevating into a worldwide war of the kind that had been so calamitous a mere ten years before).

If we take this example as representative of virtuality in general, we might conclude that a virtual *x* retains the efficiency or function of a real *x*, while manifesting these in an unfamiliar or non-customary form. Thus, the answer to the first question of why a virtual *x* counts as an *x* at all is that it retains the efficiency of the kind. This fits with a definition of virtuality formulated by Charles Sanders Peirce, "A virtual *x* (where *x* is a common noun) is something, not an *x*, which has the efficiency (virtus) of an *x*" (Peirce, 1974:

261). And the answer to the second question, what a *non-actual x* amounts to, is that the *x* may take a form that does not include the properties we ordinarily consider an *x* to have, perhaps because it is instantiated in a novel form. Hence, virtual *x* is not an actual *x*, but only if *actual* is taken to refer to the *customary means* of instantiation of some item or function. But I do not agree with Peirce that this necessarily means that a virtual *x* is "not an *x*," because the virtualization of an item or activity is a *reflexive* event that may change the extension of the category to encompass a new way of being an *x*.

To bring this discussion back to the topic at hand, a virtual store such as Amazon has "the efficiency of" a brick-and-mortar store because it achieves the same or similar function and as such is *as good as* a brick-and-mortar store for the purpose for which we find stores useful: that is, purchasing goods. But it does so in a way that is unfamiliar and non-customary (or at least was so twenty-five years ago). A virtual store is a real store because of its functional efficiency, it is just one remediated in a non-customary form. *Virtual memory* is real memory because it serves the function of actual computer memory, but it does so via non-customary means. Virtual memory occurs when a computer system lacks the capacity in Random Access Memory (RAM) to run a program, and, to compensate, the system temporarily moves parts of the required memory to addresses in physical disk storage, giving the impression that the system has more RAM than it in fact does. What is it that is virtual about virtual memory? The obvious suggestion is that virtual memory stores data not in actual RAM, but in its temporary simulation.

But now there is an important conclusion to make here: that an item has the efficiency of a precedent item, but is *non-customary*—and hence virtual—depends on the contingent facts about how some item or event has traditionally been instantiated. But I think this is exactly the right conclusion to make here. Computer memory, because of a historical quirk, could have had the design of virtual memory from the outset: and in such a case it would have been memory *simpliciter*. Similarly, in some strange world it could have been the case that online shopping had always been the norm, and that brick-and-mortar stores were at some stage invented. In this world, these weird physical stores— *imagine having to leave the house to shop!*—might count as virtual stores, achieving the function of an actual store but doing so in a non-customary way because of their instantiation in an unfamiliar form.

2.4 Virtual Remediation

Virtuality can thus be analyzed in terms of *remediation*, as the function or efficiency of an *x* being remediated in a non-customary or unfamiliar form. The connection between media and virtuality, the two key concepts in this chapter, is not accidental. When Grabarczyk and Pokropski say in the paper cited above, that "it might be best to think about virtual as a kind of fundamental remediation of one ontological domain into a different ontological domain," they come very close to the position I have developed here. However, their account is problematic because when they note that "we could define 'virtual' as 'non-physical remediations of physical objects, spaces and processes that are accessible jointly to many agents'" (2016: 27) the nature of the remediation suggested is a "non-physical" one, and so the whiff of the unreality of virtual remediation remains. To avoid this, we need only stipulate that the form of remediation is not "non-physical" but is instead in terms of a change in material medium.

Remediation is common and occurs wherever items of the same kind might be instantiated in different materials or media. But we now see that not just any act of remediation will count as a case of virtuality. As noted in the previous section, first, the remediation must be in a non-customary form. And second, the remediation must preserve the functional efficiency of the item remediated. Without the first of these conditions being met, a remediation is likely to produce a mere *replication* or *copying* of the remediated item. If the latter condition is not met, we are likely to be left with a *representation* or *imitation* of the remediated item. To see why this is so, both the customary and functional efficiency requirements of virtual remediation or virtualization need some further explanation.

At the beginning of the Iron Age in the Mediterranean, the implements of war that were traditionally cast in or beaten out of bronze—swords, spear heads, axes—were soon to be forged in iron. This remediation, comprising a key technological development of the time, gave an advantage to the military forces that adopted it, largely because of the wider availability of iron over bronze. The availability of bronze was compromised by the limited supply of its component metal tin, where iron ore, though much more difficult to process because of the higher temperatures involved, was significantly more abundant. Even if bronze swords were often individually superior, a larger army could be furnished with iron weapons, this being one of the factors that led to the growth in the size of armies in the transition from the Bronze Age to the Iron Age (Gabriel, 2006: 26). In this example, however, I doubt that many of us would be tempted to think of iron swords as *virtual swords*, because the material medium is much too close in form. Bronze and iron are the same kind of material in a broad sense, that is, solid metals, and so bronze and iron swords in practice differ very little from each other, perhaps only with respect to the technical details of their construction, their quality as swords, and their relative availability. Here remediation leaves us with a replication of the item.

Virtualization typically involves a fundamental change to how the function of the virtualized item is realized. While the overall functional efficiency of a virtual item is retained, typically, subordinating aspects of the structure and function of items are altered via the act of remediation. For example, virtual monetary systems such as cryptocurrencies are a (somewhat) effective medium of exchange, but how cryptocurrencies achieve this functional efficiency is quite different from customary forms of currency. To function as a unit of monetary exchange, items must be discrete, divisible, fungible, and scarce, and so their embodying medium must allow for the instantiation of these "sufficient differences," as Bacon puts it. Focusing on the property of scarcity, we can see how a rare metal like gold leads to a natural scarcity in supply, giving coinage based on gold an intrinsic value (that of course must still be protected from debasement). The scarcity of a dollar bill, on the other hand, is maintained by regulation and supply, and policing of forgeries. When we come to a modern virtual cryptocurrency such as Bitcoin, the scarcity that is a key part of the function of the currency, is maintained by the complicated technology of "blockchain mining." This is a very non-customary means of establishing scarcity, and indeed it is designed to avoid the customary regulative controls associated with traditional monetary systems because the designers of the "disruptive" technology of Bitcoin find these to be problematic for their purposes. This fundamental change in medium is not what is occurring in the case of the iron and bronze swords; in that case the material remediation is so minor, that the entirety of the subordinate function and structures of the sword—the sharp metal edge and point—is retained even as it is remediated.

Secondly, virtualization, or what we can now see is virtual remediation, requires a preservation of the functional efficiency of the remediated item. In a more technical sense, I have argued elsewhere that useful in understanding this sense of virtual remediation is the idea of "structural or functional isomorphism" (Tavinor, 2018: 154). "Isomorphism" is a term with applications in biology, crystallography, and mathematics that means "equal form" and is used to refer to a functional or structural correspondence between objects in different material domains. It is this isomorphism that gives remediated items the "efficiency" mentioned by Peirce. And to again draw on the case of virtual currency, it is the features of discreteness, divisibility, fungibility and scarcity that allow for an item to act as a medium of exchange; and for something to be a virtual currency it must genuinely have these features.

Without this preservation of functional or structural identity in an act of remediation, we are likely to be left with a mere representation or imitation of an item. An attempted remediation of currency that did not somehow reproduce the feature of scarcity, for example, would likely leave us with a form of currency suitable only for *play*, such as in the case of Monopoly money. A recent curious example of this is *Dogecoin*: part cryptocurrency, part meme, Dogecoin, introduced in 2013, is often suspected to be too readily available to successfully function as a unit of monetary exchange. To take a quite different example of a remediation that did not retain the functional efficiency of the original, in the 1980s, the table-top game *Dungeons and Dragons* was made into a cartoon, but the cartoon, given its non-interactive medium, did not allow for the functionality inherent in the table-top game. The cartoon was not a virtual instance of *Dungeons and Dragons*, but merely thematically borrowed from it. This probably explains why I found the cartoon so disappointing, as it did not embody the crucial qualities of the game of which I was such a fan. However, we now know that the game can be given a digital form which will preserve its interactivity, and so will count as a virtual form of *Dungeons and Dragons*. And in fact, there are multiple games that come very close to reproducing tabletop *Dungeons and Dragons* in a virtual way. The videogame series *Baldur's Gate* is set in the *Dungeons and Dragons* campaign world of the Forgotten Realms and approximates the rules of the second edition of *Advanced Dungeons and Dragons*.[6] *Dungeons and Dragons* is also often now played remotely, particularly on the online virtual tabletop website Roll20.

Virtualization thus sits between replication and representation: a virtualization of an item is not a mere reproduction of the item, including the material form along with the remediation of its structure, but neither is it just the representation of the item, lacking these qualities. In virtualization, the structure and function are retained, while the customary material instantiation is lost. This leads to the important point that in virtualization if *something is lost, what is lost is often a gain*. Features associated with the material form of an item or activity may not survive its remediation into a new material form, but this is often precisely the motivation for the act of remediation. Virtual teaching retains much of the function and structure of customary teaching, but the online nature of the activity means that copresence of teacher and student is not required, which has obvious benefits in a time of pandemic. By not functioning as a physical site of distribution, an online store can make savings from the elimination of the labor involved in traditional customer service, which is devolved into the form of a website. The loss in both cases is not an unalloyed gain for all involved parties, of course: virtual teaching and virtual stores both open the way for the *physical redundancy* of teachers and shop attendants, and we might also worry about whether activities such as virtual teaching really do preserve the critical functions of teaching in their remediation of this activity. When I come to examine VR media specifically, the issue of what is gained and lost in remediation is going to have consequences for my account of realism in those media.

We have become accustomed to associating virtualization with computers, but this kind of non-customary functional remediation is not limited to digital media. But while it exists outside of computers—we can now see that telephones really are virtual devices because of how they preserve the functional efficiency of conversation in a non-customary way—virtuality is now strongly (though contingently) linked with computers. Why are computers so apt for virtualization? The obvious answer here is that, as noted earlier, computers are *imitation machines*. A computer is a machine that can be instructed to carry out logical and arithmetic operations, to solve problems or produce results. When it is connected to input and output devices, such a machine is capable of manifesting the functional efficiency previously associated with a vast array of everyday objects, processes, or activities. The contingent connection between computation and virtualization exists because computers are apt, in virtue of their very nature, to remediate functional artefacts and activities. Because virtuality extends beyond the realm of the computational, we might refer to the specific class of *computational virtuality*. This class includes those cases of virtual remediation that exploit the functional and structural potential of computers; or, simply, virtuality in computational media. And while the notion of virtuality clearly exists beyond computers, it is within this domain that virtualization is most stark and potentially puzzling, because of the fantastic potential inherent in these new machines and the radical effects they are having on our lives and culture.

A final point here is that virtual items have the habit, after a suitable period of familiarization, of becoming actual instances of the things they at first only virtually manifest. Virtualization, I noted in the previous section, is a reflexive process. Telephones are a virtual technology, but this is an unfamiliar way to describe telephone conversation because after over a century of their existence, we have come to think of telephones as a customary way of speaking to people. Virtual stores are now just stores; and indeed, just what it is to be a store has now changed to reflect the prevalence of online shopping. The customary mode of achieving some end, changes, of course, with custom. Furthermore, the precise nature of the remediation involved in virtuality is variable and historically contingent, depending on the kind of media available at a given time, and our expectations about what is a customary way of achieving a goal. This is not an insignificant fact, because it says something deep about my view of virtuality, and hence, about virtual reality media. Virtuality is not a new thing, much less a symptom of our current technological predicament, because aspects of it, perhaps primitive or partial, have long characterized cultural development.

2.5 The Remediation of What?

With a general view of virtuality before us, let us now move on to examine how virtualization manifests in the narrower class of virtual reality media that are the specific focus of this book. So, if a virtual medium is a means of preserving or instantiating a function or structure in a non-customary way, what is the function or structure that the VR media under discussion here preserve? What kind of remediation is involved in so-called *virtual reality*? What is the functional efficiency instilled in the cases I have been exploring above, such as in my vulnerable and terrified experience of the world of *Resident Evil*?

It should be clear enough that the principal thing that is being remediated in such cases is *an agent's experiential and interactive dealings with a world*. Inspecting the examples already introduced, and others, substantiates this view. In *Resident Evil VII*, the player sees and hears the world *as though* through the eyes and ears of the character within the world, and they may encounter and manipulate objects, such as switching on TVs or opening doors, *as though* they were an agent in the world depicted by the game. In *Tumble VR* (2016), a VR based puzzle game on the PS4, the player can inspect the puzzle pieces before them, and may pick them up, stack and balance the different geometrical shapes, and even have a tactile sense of feeling the pieces, as though one were solving a physical puzzle actually before them. In *Tilt Brush*, an open-source 3D virtual reality painting application, formerly distributed by Google, the player is visually situated within an abstract 3D space and given a collection of virtual artistic tools such as paint brushes, color wheels, backdrops, and lighting, with which they can fill that space with configurations of color and shape to create a kind of virtual sculpture.

Earlier in this chapter, we found that the notion of a medium is used in a number of crucial contexts including the basic physicalist sense of a material in which items are instantiated or through which physical forces or movement are transmitted, the mentalistic parallel of this as a substance or structure via which sensory impressions, experiences or ideas are received or conveyed, and the agential context where a material object, act, or process is conceived as a means to some end. The experiential and agential senses seem to be an obvious focus in the current context given that it is the presentation of apparent interactive sensory environments—*virtual spaces or realities*—that generate the most striking aspects of VR media, and that prompt their consideration as a uniquely realistic medium. My focus here, then, is the sense in which VR media convey the features we ordinarily associate with natural sensory or perceptual experience of a spatial environment, allowing users sensory and agential access to an apparent virtual world around them; moreover, the way in which virtual media frequently bear a striking functional and structural correspondence with ordinary experience of, and action in, the world. Hence, the functional or structure preserved that makes these media *virtual*, is of an epistemic and agential orientation on a space, normally attributed to our native perceptual and behavioural access to the world, but in this case embodied in depictive and interactive media.

Here, then, we have the beginnings of an answer to the question of why *immersion, interactivity*, and *computability*, identified in both Heim's and Chalmers' definitions, lie at the heart of VR media. Immersion, though we will find the concept is a problematic one, refers to the experiential nature of VR, by referring in a metaphorical sense to the potential in VR for users to adopt an alternative experiential perspective. Interactivity refers to our apparent agency in VR, that is, the impression that VR media remediate the native ability to act and change the environment in which one exists. Computability refers to a technological medium that has realized VR media as a viable technology in the last quarter century. The concepts of immersion, interactivity, and computability are all going to need further attention and clarification, but the combination of these features genuinely does help with identifying what is remediated in VR, and the technological medium that is the physical substrate allowing for this remediation of experience and agency.

However, this view immediately puts my account in opposition to what seems to be an alternative interpretation of the remediating nature of VR. This is the view that it is the *objects* depicted in VR worlds that are remediated by the technology. When Chalmers notes that "virtual objects are real objects" one might naturally understand this as the claim that virtual reality technology remediates the objects of our experience, giving them a kind of virtual (but also real) existence (2017: 309). Similarly, when Brey claims that "virtual apples simulate or imitate real apples" it seems obvious that he has in mind that a virtual apple might be considered as a remediation of a real apple (2014: 43). Hence, we have at least two accounts of VR media. First, is the account that I will favor, that VR media allow for the remediation of experience and agency; and second, the position that it is the objects and worlds experienced or interacted with that are remediated in VR. That these are different and competing positions, and just why I favor the former position, is going to need a good deal of argument in what follows. This will begin in the next chapter with a technical explanation of my own favored view.

Notes

1 The owners of the *View-Master* brand, toy maker Mattel, have recently made efforts to produce headsets with VR functionality to be marketed under the brand.
2 This is also related to the sense of "medium" as a person able to receive "spiritual" impressions and count as the middle ground through which such impressions might pass, perhaps from the dead to the living.
3 I am obviously passing over a lot of detail here, as this section could have involved a long digression on the nature of modern computers, and their conceptual foundation in Turing machines. For an interesting discussion of the details, the reader might like to consult Copeland, 2005.
4 We should be careful not to become confused by the extra complication here that programs, code, and other computer artefacts are themselves "substrate independent" in nature (that is, they can move between different hardware implementations). But while a specific program or algorithm might seem "non-physical" in being functionally defined, the medium as a whole is no less physical than media such as newsprint, paintings, photography. Computers do not work by magic.
5 An often-cited precedent for this view is Gilles Deleuze (2002), who argues that the informative contrast for the term "virtual" is "actual," rather than "real." Unfortunately, Deleuze's work is not particularly helpful in clarifying the issues here as his paper is marred by repetitive, overstretched metaphors and jargon appropriated from physics (mixed, it seems from electrical engineering, nuclear physics, optics) and is also epigrammatic and vague. The translator of his work, Elliot Ross Albert, suspects the work is an incomplete early draft.
6 This is more complicated than I let on here because it is not clear just which features of a game must be preserved for a reproduction to count as a remediation. For an account of the limits on the possible remediation of games, see (Bartel, 2018).

3 The Virtual Remediation of Spatial Experience

3.1 The Depiction of Space

It is credible that VR media such as stereoscopic headsets and virtual audio are intended to convey the features we ordinarily associate with natural sensory or perceptual experience of a spatial environment. Thus understood, we can now inspect the technical elements of VR media via a discussion of how the experience of space is remediated by virtual technologies. This chapter will explore in a detailed way, not only the current state of the art of virtual media, but also how current virtual media have grown out of and extended the capacities of earlier forms. It might be noted here that the technical detail of this chapter is not always entirely philosophical in nature; but this technical content is needed to properly ground the philosophical arguments to follow.[1]

In terms of vision, the media device typically associated with VR is a stereoscopic headset, but we should note that this is merely the currently most prevalent way of depicting visual spaces in VR and that other means such as the wrap-around parabolic projection screens used in theme park "dark rides" such as Universal City's *Harry Potter and the Forbidden Journey* can also give a visual sense of spatial presence and immersion. What these visual virtual media share is the depiction of the user's apparent position within the depicted space. But modern digital media do not comprise the first attempts to remediate visual spatial experience. To see this, we need a short historical digression to understand several precedents of the virtualization of spatial experience.

Many traditions and genres of picturing have in the past attempted to depict the spatial arrangement of objects and scenes in a naturalistic way. In the High Renaissance, in paintings such as Raphael's *School of Athens* (1509–1511), the sense of spatiality and the arrangement of figures within depictive space were a principal aim and are considered central to their great artistic achievement. Raphael's famous fresco, painted on the walls of the reception rooms of the Apostolic Palace in the Vatican, gives a strong impression of a high-vaulted architectural space projecting away from the viewer. Raphael depicts as its central figures Plato and Aristotle, the former gesturing upward to the heavens or the forms, and the latter downward and forward out of the scene. Plato's gesture is confined to the picture space, as

DOI: 10.4324/9781003107644-3

he points upwards, his forearm parallel to the painted surface. But Aristotle seems to reach outside of the picture, his foreshortened arm projecting towards us the viewer. The space of the scene is thus not only a virtuosic and naturalistic rendering of spatial experience but is also critical to understanding the work because of how it engages the viewer in the action of the scene depicted, giving the impression that they too are figures within the picture's implied world of learning, and inheritors of the Classical tradition of philosophy.

Several different painterly techniques were used to give this impression of picture space in the works of the Renaissance. Most basically perhaps is the arrangement, relative sizing, and occlusion of figures within a depicted space. In Raphael's painting, the figures are overlayed on one another so that the viewer receives a very natural impression of who is at the front of the scene and who is towards the back; this impression is strengthened by the sizing of the figures, and the viewer's prior expectations about the relative uniformity of the size of people. These are very basic features that arise when one views a natural scene from a defined position, but it might be noted that some pictorial traditions do not adhere to even these "obvious" ways of arranging figures on a surface to give an impression of space, and their use does in fact comprise a discovery, if not an advancement, of depictive form.

Another example is the technique of atmospheric or aerial perspective (*prospettiva aerea*) which can give a scene a sense of depth, or as receding from the viewer, a famous example being Leonardo da Vinci's *Virgin of the Rocks* (1495–1508). In atmospheric perspective, objects in the distance are painted impressionistically and with a desaturated bluish tinge to replicate the effects of light being scattered by particles of moisture and dust as it travels over long distances. Another painterly technique, *chiaroscuro*—literally, "light-dark"—is a method of using tonal contrasts to give volume or three dimensionality to objects, and again was used by Renaissance artists such as Leonardo da Vinci and in the baroque works of Caravaggio to give a sense of space. These methods of giving an impression of picture space simulate the behavior and appearance of light when it interacts with objects and environments, rather than directly conveying a sense of the geometry of the pictured space through the configuration of lines and shapes. However, a large part of the success of Raphael's fresco relies on it pursuing the latter via its masterful use of *linear perspective*, a technique that has come to bear much of the theoretical burden in discussions of painterly naturalism.

A crucial development in the technique of linear perspective—and critical to the masterworks of the High Renaissance—were experiments performed by Filippo Brunelleschi in Florence in 1413. Standing adjacent to the Cattedrale di Santa Maria del Fiore, for which he would eventually engineer the dome, and employing a panel depicting the Florentine Baptistery painted in his newly developed linear perspective style, Brunelleschi contrived to show how the technique reproduced the geometry of the building in a pictorial medium. Brunelleschi's early biographer, Manetti, describes how the painter,

> made a hole in the painted panel at that point in the temple of San Giovanni which is directly opposite the eye of anyone positioned inside the central portal of Santa Maria del Fiore [...] The hole was as tiny as a lentil bean on the painted side and it widened conically like a woman's straw hat to about the circumference of a ducat, or a bit more, on the reverse side. Whoever wanted to look at it was required to place his eye on the reverse side where the hole was large, and while bringing the hole up to his eye with one hand, to hold a flat mirror with the other hand in such a way that the painting would be reflected in it.
>
> (Manetti, 1970: 44)

Brunelleschi's intention was to directly compare his painting with the qualities of natural vision to show how his technique rendered the geometry of natural vision in a potentially illusionistic way. The cultural impact of the discovery was rapid, with the technique quickly being adopted by his contemporaries Donatello and Masaccio (Edgerton, 1973: 172).

Leon Battista Alberti would go beyond these experiments in his codification of linear perspective in the treatise *De Pictura* (1435/1973). Alberti designed a system of perspective—a *perspectiva artificialis*—that treated painting in terms of an imaginary "picture plane" suspended between the viewer and the depicted scene and allowing for the organization of perceptual space by charting the geometry of light rays. This plane acts as the base of a tilted pyramid, where the peak of the pyramid extending away from the viewer represents the vanishing point of light rays. Employing a horizon line that runs through the vanishing point and orthogonal lines that run perpendicular to the picture plane, the bottom side of the tilted pyramid can be depicted as a foreshortened tiled floor. This is called the "plan and elevation" method of perspective (*costruzione legitimma*) and its use allows for figures and objects to be appropriately scaled and positioned in the picture space. Of course, all these lines are scribed on a flat surface, merely giving the *impression* of the spatial configurations described above, and this, indeed, is the magical effect of the technique.

Alberti's refinement of Brunelleschi's attempt to replicate the perceived qualities of visual scenes in a visual artistic medium was an important technological advancement because the perspectival techniques used by artists stretching at least as far back as the Romans were intuitive, accidental or less than theoretically informed, in comparison to the coherent theory seen in Alberti's plan and elevation method. Alberti's technological contribution for conveying convincing picture space was applied by contemporary artists and was influential on the future artistic depiction of space, famously in Pietro Perugino's fresco in the Sistine Chapel (1481–82) and then in the works of the High Renaissance, such as Raphael's *School of Athens*. But beyond this, the technique profoundly reconfigured the depiction of space in European painting, and perhaps even influenced the rational conception of space in the scientific revolution that would follow the Renaissance. Samuel Edgerton (2009), for example, argues that the development of linear perspective, considered as a part of the science of optics in its own time, was a key to the development of the scientific image of the world. The philosopher John Hyman notes that linear perspective had a "theoretical" effect on science, because space could now be conceived in rational and principled terms, and with the development of analytic geometry, the spatiality of the world could be conveyed "in three dimensions in a uniform and comprehensive geometrical scheme" (Hyman, 2006: 211).

Moreover, because it represents space in terms of a vanishing point on the horizon, thus giving the impression of the space projecting away from the viewer, the technique represents a *perspective* on space, or the impression of *looking into* the picture's world. The technique implies the presence and even position of the viewer by representing the spatial experience of a subject, and so the viewer seems to become a more prominent part of the picture's experience and interpretation. Linear perspective gives the impression that the viewer stands not just before the picture, but *before the scene pictured*. This was what allowed Raphael to place the viewer—perhaps most of all his sponsor, Pope Julius II—in the same space as Plato and Aristotle. This technique also opened the way, centuries later, for the spatial effect of VR media.

Even so, paintings or engravings in linear perspective, even those virtuoso works of the High Renaissance, seem flat in comparison to natural experience. While of course this is partly due to that fact that these works do not disavow their nature as surfaces—in artistic paintings and drawings a large amount of the interest is in *how* the flat surface generates the spatiality—it could also be attributable to the failings of linear perspective as a remediation of spatial experience. There are notorious problems with linear perspective if it is treated as a "copy" of natural visual spatial perception. In his critique of pictorial naturalism, Nelson Goodman identifies his target as the idea that linear perspective was a "long stride forward in realistic depiction," (1976: 10) a position which he understands to claim that,

> A picture drawn in correct perspective will, under specified conditions, deliver to the eye a bundle of light rays matching that delivered by the object itself. This matching is a purely objective matter, measurable by instruments. And such matching constitutes fidelity of representation; for since light rays are all that the eye can receive from either picture or object, identity in patterns of light rays must constitute identity of appearance.
>
> (Goodman, 1976: 11)

His argument against this position comes in two parts. First, we must grant certain constraints if perspective is to be seen as faithfully copying the appearance of objects—that is, that the picture be viewed "through a peephole, face on, from a certain distance, with one eye closed and the other motionless"—and already these constraints destroy much of what it is to naturally perceive an object because they specify "conditions of observation [that] are grossly abnormal" (13). Second, even if we accept these constraints as unproblematic, the geometry of "bundles of light rays" so furnished by the "laws" of linear perspective do not always convey scenes that are judged to be realistic. Indeed, they are often rejected in favor of conventional ways of depicting items. He cites as an example a geometrical configuration produced by the artist Paul Klee in his *Pedagogical Sketchbook* (1953) which under the laws of linear perspective render what should be seen when standing alternately before a floor projecting away, or a façade of a building projecting upwards (Goodman, 1976: 16–18). When interpreted as a floor, the configuration looks realistic to most viewers, the lines of floor converging as would railway tracks depicted in linear perspective. But conceived as a façade the edges of the building seem to unnaturally converge toward the top, where convention—including that employed to depict the architectural forms seen in Raphael's *School of Athens*—would have them parallel. Klee claims that we psychologically reject the picture as a façade because "every creature, in order to preserve his balance, insists on seeing actual verticals projected as such" (1953: 41). There is also more recent experimental evidence that linear perspective is not judged as a particularly realistic depiction of space compared to alternative modes of depiction (Burleigh, Pepperell and Ruta, 2018).

There are also many ways in which linear perspective fails to capture the character of natural space. Linear perspective leads to distortions, particularly at the edges of images where objects become unnaturally stretched (Todorovic, 2009). It also typically depicts a limited field of vision compared to natural vision. These problems discount the idea that there can be purely geometrical schemes of spatial depiction that match native vision. But even if it did match the geometry of natural vision in a way that was judged to be realistic, linear perspective would not by itself be sufficient for the rendering of convincing visual scenes. In native vision, occlusion, light and shade, and textural gradients all add additional cues to the spatial depth of visual scenes, and these cues have frequently been incorporated in traditional art (Kubovy, 1986). Lacking strict linear perspective, Jan van Eyck's *Arnolfini Portrait* employs the light source provided by a window to give an effective sense of volume to the represented space; the brightly lit space beyond the two standing figures contrasts with some of the darker foreground areas. And as already noted, another spatial cue in large landscape scenes is the diffusion of light sources through the atmosphere, and the bluish desaturated appearance this gives to distant objects. This shows that depictive naturalism is not merely a matter of rendering spatial geometry, but that it also relies on providing the kinds of visual cues that our native vision employs to register the spatial depth of natural scenes.

And of course, these attempts at realistic depiction failed in another radical respect: the action in the pictures described above is frozen and fossilized before us, fixed at the time of the picturing act. In native vision the eyes and head are always moving, and the static nature of painting cannot reproduce this movement and its effects on the apparent spatiality of the scene. And more importantly, the scenes depicted are themselves static. There is no movement in Raphael's *School of Athens*, and at most it can be *implied* by the bodily poses of the gathered philosophers. Contrapposto, a sculptural technique developed by the Ancient Greeks, is where a human figure is depicted with most of its weight on one foot. Though the figure does not move, the pose gives the impression of tension and the potential for movement, and it was widely employed in the Renaissance (the figure of Aristotle in *The School of Athens* is depicted in contrapposto, for example).

While the lack of actual movement cannot really be called a *failing* of the pictures—the static content is to be considered a standard feature of such paintings (Walton, 1970)—it is a severe limitation for their potential as an accurate remediation of spatial experience. In the years since the High Renaissance, the depiction of 3D space on a 2D surface has been either embraced or repudiated as with artistic fashion the interest and importance assigned to this feature of depiction by artists and critics fluctuated. In some later traditions, including analytic cubism, the space of the picture was fragmented, flattened; and in others, such as abstract expressionism, the painting seems entirely unconcerned with the depiction of space in the scene.

But even given these limitations, the mere interest in conveying the experience of space is already, it seems to me, an intended virtual depiction of reality. In terms of the theory presented in this book, we could understand Brunelleschi as seeking to remediate the experience of space in pictorial form. Alberti's work went further and codified the pictorial remediation of spatial experience. Here then, we can see in traditional art there are multiple precedents for the kind of spatial depiction seen in VR. And already in 1605, setting out the technique of linear space for an audience of artists, Hans Vredeman de Vries was producing engravings that are reminiscent of the wireframe models that would eventually be employed for the purposes of realizing virtual worlds. Through such pictures, we receive the impression of looking into a space. These traditional painterly techniques may even allow for the experience of the feeling of *presence* that is so frequently associated with VR: standing before a large *trompe l'oeil* work such as Samuel van Hoogstraten's 1662 painting, *A View Through a House*, one really does get the sense of occupying the space of the picture. While the flow of natural perceptual experience is not conveyed, the configuration of the painting's surface gives the vivid impression that one could almost step into the space represented.

Artists have thus long been interested in conveying a sense of space and conceiving this interest as something like the attempt to remediate our perceptual experience of the actual world. French film critic and theorist André Bazin considered the impact of linear perspective on Western painting was such that it, "made it possible for artists to create the illusion of three-dimensional space in which objects could be placed the way they would if we perceived them directly" (2009: 5). If we consider VR media to aim for the remediation of spatial experience, we can recognize pictures in linear perspective as an important precedent for this aim. And when seen in light of these earlier attempts to remediate the experience of the world, the development of VR in many ways might seem like a return to the naturalistic aspirations of this art. It is, of course, an open question whether VR media really do meet this aim of remediating the experience of space, which is something to be addressed in Chapter Six. But in the following sections of this chapter, we will discover that computational virtuality allows VR media to solve many of the problems associated with the remediation of spatial experience encountered by linear perspective, because of how it can expand on the veracity of the *perspectiva artificialis*.

3.2 3D Graphical Imagery

While Brunelleschi's and Alberti's picture planes and the works of the High Renaissance give an impression of space within the picture, they are also spatially problematic. Historical attempts to remediate the experience of space in terms of picture space were usually flawed or incomplete. And as already noted, eventually there was a certain amount of skepticism that any such remediation was possible at all. I will argue here that VR media can be understood as something of a "solution" to these issues because plausibly, the represented spaces of VR are more spatially precise and do more to realize the viewer's perspective on a visual scene. They are also obviously much more dynamic and interactive than these earlier media could be.

What are the specific technical developments that allow for these solutions? VR's advancement over previous depictive media depends on three principal parts: its use of a dynamic 3D pictorial medium; the tracking of a depictive perspective within this medium; and the display of this perspective by a viewing device, usually stereoscopic in nature, that allows for egocentric viewing. The following three sections explore these issues in turn.

First then, let us turn our attention to the techniques of *image production* involved in visual VR media. So that we can properly understand these developments, I will begin with a potted history of the technology of image making itself. The technical production of images has undergone numerous developments and transformations over the course of the history of the arts and sciences. Images need not be pictures of course: the upside-down images on the retina of the eye are not pictures of what we see, they are the visual registration of light waves that partly constitute the activity of seeing. Crucial in all images, and the narrower range of these that we appropriately call pictures, is a distribution or configuration of sensible visual qualities on a surface, array, or plane. In the case of picturing, this configuration comprises a marked surface, intentionally produced for its representational capacity. The material qualities of the surfaces of such representations are varied: the walls of a prehistoric cave may be daubed in rudimentary pigments such as ochre and charcoal to depict stylized horses, stags and cattle; a fresh lime wall might be decorated with water-soluble pigment, subtly applied to fresh plaster to represent religious scenes; light can be focused via lenses on a light sensitive emulsion carried on a substrate such as cellulose acetate or glass, and then chemically developed into a image to be carried on paper as a photographic print; in a cathode ray tube, beams of electrons can be shot through a vacuum to scan across a phosphorescent screen, generating the appearance of moving images, and so depicting the adventures of Buck Rogers in the 25th Century.

In all these cases, the viewer sees the configuration of the surface—in some cases static, in others animated and moving—and perceives the content of what is being depicted. They do so both by perceiving the bare spatial configuration— that is, that there is a space, and that objects and figures may occupy that space in some more or less spatially definite way—and what this space represents—an auroch, a nativity scene, a recently married couple, Buck Rogers defending the Earth from invaders. The surface configuration comprises shapes, borders, but perhaps at its most basic is a configuration of colors through which we see these. For Hyman, pictures most fundamentally "consist of colors [including the achromatic colors black, grey, white] distributed on a plane" (2006: xvii). Explaining just how we perceive the configuration of colors as comprising borders and shapes, and how these shapes and borders give the impression of a space occupied by objects and figures, is likely to draw on basic facts about the human visual system and its ability to discern spatial reality. The details of this visual engagement will be a crucial part of any full explanation of VR media, and I will need to return to certain features of the visual system in the following discussion of the experience of the egocentric pictures of VR.

For a long while, hand-made pictures predominated. In a simple sketch on paper, I might use a pencil to draw a house with a very basic, almost symbolic rendering, as a set of geometric primitives: a square for the front of the house; a triangle atop this for the roof, and rectangles for the windows and chimney. In Raphael's rather more accomplished frescos in the Apostolic apartments, it was his intention, having been commissioned by Pope Julius II, to depict scenes suitable to conveying the Church's relationship to philosophy, jurisprudence, theology, and poetry in a way that reconciled Christianity with Antiquity. For this purpose, the *buon fresco* medium and various depictive techniques including linear perspective, were chosen. Raphael employed these media and techniques to physically place perceptible marks on a surface that give a distinct sense of space within the picture, and objects and figures existing in that space. In hand-drawn pictures, the use of these depictive media in the laying down of the surface involves the judgment and skill of the image maker, but this judgment can be augmented by rules or principles of picturing, of which the linear perspective codified by Alberti was just one form. In this case, rudimentary mathematic and scientific principles can be employed to configure the surface in the hope of giving a precise spatial impression. Nevertheless, in both my rudimentary sketch of a house, and Raphael's painting, the judgments of the image maker intervene between the content of the picture, and how this is rendered on the surface, and the resulting image is produced by a physical act that renders the object visible to us.

Science and technology have always influenced pictorial media, and in the 19th Century, scientific advances led to the development of photography and similar forms of image production. These means of producing pictures—of laying down a pictorial surface—are principally mechanical and physico-chemical, usually involving two parts: a camera used in the registration of the image of the scene being pictured, and a means of fixing this image on a pictorial surface. A *camera obscūra*—literally "dark chamber"—comprises a darkened room with an aperture on one wall and a screen opposite. Light enters the aperture, which may be a small hole or lens, and is propagated in straight lines to the opposite wall where it registers as an image—inverted, and back to front—as the light waves are absorbed and reflected by the surface. *Camera obscūras* have precedents in the natural world, both in the physical environment in the form of naturally occurring pinhole cameras, and in the biological world in the form of eyes. But as designed artefacts, they have existed since antiquity. They were studied by the Arab philosopher, mathematician, and scientist Ibn al-Haytham in the 11th Century, leading to his influential optical conjectures about the nature of vision, published in his *Kitab al-Manathir (Book of Optics)*.

The images and apparent spatial configurations produced by a *camera obscūra* owe to the physical principles on which the artefact operates, and thus many of the features employed for configuring a 2D surface to give an impression of 3D space, including occlusion, relative sizing, and the apparent vanishing point of linear perspective, can arise quite automatically out of the mechanical operation of a camera (when, of course, its operating parts are designed to accurately produce these spatial elements). It is credible then that the camera operates to *capture* on a 2D surface the genuine elements of reality that are before its lens. Indeed, it was partly based on an understanding of the *camera obscūra*—distributed into Italy in the 14th century via an Italian translation of *De Aspectibus*, itself a Latin translation of Ibn al-Haytham's optical treatise (Smith, 2001)—that Lorenzo Ghiberti, Brunelleschi and Alberti were able to develop the technique of linear perspective, and also to justify the naturalism of the method (Al-Khalili, 2015). The *camera obscūra* underwent a continual development after this time, including in a scientific paper presented by the prolific scientist Robert Hooke to the Royal Society in 1694. Hooke proposed the camera as a tool to aid artists in the production of naturalistic pictures, and it was perhaps used for this purpose by artists such as Vermeer, though this is controversial (Steadman, 2001).

The principal developments of 19th Century photography involved various technologies being designed to register or "fix" the image produced by such cameras to become durable and portable. In the daguerreotype, a popular early method of photography, a polished silver-plated copper sheet is exposed to halogen fumes which induce the plate to become light sensitive. When exposed to the focused image of a camera the image is captured on the plate and this is then fixed by further chemical treatment. Daguerreotype photography was expensive and technically difficult to employ, and so was quickly superseded by film photography, which used the intervening step of an initial latent image, or negative, to produce potentially multiple positive copies of the image captured. Film would come to dominate photography, a position it would only lose at the end of the 20th Century when it was largely replaced by digital modes of capturing, displaying, and storing pictorial imagery.

These mechanical means of image-making, and particularly their prominent differences to earlier forms, led to no lack of theoretical and philosophical debate. One important point of debate is how the role of intentions and creativity seen in previous kinds of image-making such as sketching and painting, seem disrupted in the case of mechanically produced images (Bazin, 2009: 6–9) and the resulting question of whether photography can count as a representational art form at all (Scruton, 1983). But this same automaticity or mechanical operation of the camera might be thought to add to it some special epistemological privilege or truthfulness, an issue that has concerned many philosophers when they have addressed the aesthetics of photography.[2]

Photography had an enormous impact on spatial depiction—and ultimately on the potential for VR—in large part because it paved the way for the art of the moving image. By capturing and collecting a sequence of images, produced either by the photography of live scenes or by photographing sequences of hand-drawn pictures or "cels," and projecting these in quick succession onto a screen, photography could be employed to produce animated scenes. These techniques allowed for the rise of the cinematic genres of live action and animated film.

The development of moving pictures and animation adds a great deal to the depiction of space because the movement of configurations on a 2D pictorial plane allows for what Erwin Panofsky calls in an early influential essay on cinema, the "*dynamization of space*" (1947). Object occlusion and relative sizing are already spatial features of static depiction, but moving objects allow for objects to move in and out of occlusion, strengthening the impression of the spatial positioning of figures. Objects can also move closer to and further from the camera, changing their relative size in the picture space. Motion parallax can occur when a live action camera moves through a scene, and objects close to the camera move more quickly than those further away. This effect can also be contrived in cel animation such as in several classic Disney films, where multiple transparent layers depicting planes at different apparent distances from the camera can be animated to move at different speeds across the picture plane to give the impression of parallax. And it should be noted that while such animation does involve a camera in capturing the animated cels, the motion of the camera *through* the depicted space in such a case is merely apparent, in being an artefact of the sequence of images. All of these techniques allow viewers to make judgements about the relative position and motion of objects in visual space, and the meaningful and expressive effects of cinema that arise from this spatiality.

Against the technical background of this potted history of image and picture production, we can now understand the developments that have led to the picturing processes at the basis of virtual reality media. Simplifying things somewhat, we see above that moving pictures usually relied on one of two methods of depicting dynamic space: either simply placing a movie camera within a space and filming the action, or by producing animated scenes one image at a time by photographing hand drawn pictures. In the first, the production of the spatiality of the scene is achieved automatically by moving an actual camera though a real scene, and in the second the production of spatiality is more akin to the depiction of space in hand-drawn pictures via the intentional production of marks on a pictorial surface.[3] We might think that VR media occupy a place halfway between these methods of rendering moving space on a pictorial surface. This is because, like live action film, VR usually involves something like "placing a camera" within an existing scene and having the apparent spatiality of the picture arise automatically from this camera placement. But like traditional animation, the space in which the camera is placed is itself typically the creation of artists working in a depictive medium.

The depictive medium employed in VR is popularly known as computer generated imagery (CGI). CGI is perhaps the key technological development in image-making in recent times and it has radically altered how we produce and consume many kinds of visual media. The technique has surpassed traditional modes of image-making such as hand-drawn cel animation and has revolutionized the production of traditional visual media including television and film. It has allowed for, and developed in concert with, entirely new cultural forms such as videogames. It has also led to some surprising pictorial trends, such as digital image manipulation and the "deepfake" images that might provoke in us some small amount of worry about the epistemological future of picturing. This technology is so specialized and so fast developing, that I cannot do much but write about it in quite abstract terms in the following paragraphs, concentrating on those aspects of computer graphics that give rise to the impression of spatiality in VR media.

CGI refers quite liberally to graphical elements, pictures or animations produced on a computer, from very simple static 2D bitmap images to sophisticated and interactive virtual worlds. The significant CGI development for VR media is 3D graphics technology, and specifically, 3D animation. Ultimately, it is 3D animation which provides the dynamic graphical space encountered in VR media. At the basis of this technology is the mathematical specification of graphical objects, defined first in terms of spatial vertices, edges, and faces, and the polygons—geometrical plan figures—constructed from these elements. These polygons are the building blocks of the 3D digital models that populate the graphical environments of virtual worlds, something that seems strangely reminiscent of the Platonic solids discussed in the *Timaeus*. In that dialogue, according to Timaeus, each of the four basic elements—earth, air, fire, and water—are made up of particles of a specific geometrical shape: cubes, octahedrons, tetrahedrons, and icosahedrons respectively. These particles are themselves made up of polygons, specifically, squares and triangles.

But the philosopher of real relevance here is Descartes, who in the development of the Cartesian coordinates of analytical geometry made it possible to define points, lines, and shapes in abstract space using mathematical formulae. Given their mathematical basis, these abstract spatial configurations—and their geometrical modification—are also definable in the operation of computer code, and so they can be processed via a computer program to be displayed in graphical form on a screen. In that graphical context, the mathematically defined primitives can be combined into complex arrangements—a polygon mesh—to represent an endless variety of spatial configurations. As early as 1972, these principles would allow the future co-founder of Pixar and head of Walt Disney Animation Studios, Edwin Catmull, to digitize his hand in polygonal form to be rendered as an animated graphical object on a 2D screen, perhaps the very first instance of 3D animation. In the early days of computer animation, objects were made out of a relatively limited number of polygons, and so had a characteristically pointy and unnatural appearance. Advances in the art and technology of computer graphics has exploded the number of polygon faces used to sculpt graphical objects, contributing a great deal more naturalism to the objects thus depicted.

These mathematical models are the graphical basis of the virtual worlds that we encounter in 3D videogames and VR applications. But crucial to the impression of their spatiality is that the objects are placed relative to a virtual light source, allowing for the production of shadows and the differential reflection of light from the various polygonal surfaces that make up the graphical model. I noted earlier that the pictorial representation of light sources and the associated technique of chiaroscuro were a frequent part of the spatiality of traditional modes of picturing, allowing pictures to depict borders, shadows, and gradations of color that give the impression of three-dimensional objects on a two-dimensional plane. Thus, computer graphics shares with these earlier forms of picturing the understanding that the imitation of the behavior of light is essential to the effective rendering of space. When polygonal models are overlaid with texture maps and graphically realized via an increasingly sophisticated collection of rendering techniques—some of which further simulate the behavior of light, such as the CGI technique of "ray tracing"—the impression of volume and solidity of such graphical models can be incredibly convincing.

This impression of spatiality arises, obviously, only when an abstract computational model is rendered in a perceptible form, be it on a screen or some other medium. Prior to these models being rendered on a surface, they exist as data structures embodied in computer code. The realization of these models as visible objects requires the production of a picture plane, and hence, the use of a *virtual camera* to capture a perspective on the potential scene. However, some potential confusion may exist in this term. Abstractly, a virtual camera comprises a mathematically defined position and orientation in graphical space. More technically, a virtual camera is a functional unit within 3D computer graphics applications used for producing 2D displays from 3D models, or the "viewing transformations [...] that place objects from the virtual world onto a virtual screen, based on the particular viewpoint of a virtual eye" (LaValle, 2019: 86). As we have found, a camera is fundamentally a means of producing images, and this is exactly what a virtual camera does in the context of computer graphics, even though the method is computational and algorithmic, rather than mechanical. We might think then, that given the analysis of virtuality above, a virtual camera is simply a real camera, though a non-customary or unfamiliar one by comprising an algorithm.

When the virtual camera is thus "placed" within the modelled environment and this perspective is rendered on a screen, a pictorial surface is produced that gives an impression of spatiality. Many of the sources of this apparent spatiality are those encountered in traditional modes of picture-making, including the techniques of perspective, lighting, and occlusion, and the arrangement and relative sizing of figures detailed earlier in this chapter. So, depending on where the camera is placed in a scene, different parts of that scene will be occluded, the sizing of the figures in the scene will change given their position relative to the camera, and the space of the picture will project away from the apparent vantage point of the camera.

The resulting pictures and their spatial techniques would be recognized as such by Brunelleschi, Alberti, and Raphael. But are these computer created spaces built on the principles of linear perspective? That is, would Alberti recognize his technique specifically in these graphical representations? This very much depends on a given application of computer graphics. While these environments can be considered as Euclidian spaces specified in terms of Cartesian coordinates, and many cinematic applications, games and apps do appear to apply principles of linear perspective in the depiction of these spaces, frequently this is not the case. In a videogame such as *Animal Crossing: New Horizons* (Nintendo, 2020), the stylized depicted world does not seem like one where the "laws" of linear perspective always apply: the environment that the player encounters on their island curves unnaturally away to the horizon, giving an impression of a curvilinear space, even as indoors, linear perspective more consistently applies. Or it could simply be that, outside, the *world* of *Animal Crossing* is a strange non-Euclidean space! Admittedly, this example is ambiguous, but other CGI applications clearly adopt alternative modes of spatial depiction as their guiding principle, such as the stylized and chaotic modes of spatial configuration found in the wonderful VR puzzle game *Ghost Giant* (Zoink Games, 2019). The lesson here, perhaps, is that CGI, like other modes of picturing, need not have as its intention the naturalistic or realistic depiction of space.

The spatiality of CGI imagery is also fundamentally associated with animation and movement in its graphical environments. While it can produce still images, computer graphics, like live action and animation, is typically a means of producing moving images. In this role it has increasingly displaced some traditional forms of film production, so that now many putatively "live action movies" contain extensive use of CGI, and the entire medium of hand drawn animated movies has largely been replaced by CGI production in studios such as Pixar. There are various ways of producing animated scenes, but all involve the manipulation of the mathematically defined 3D models and the perspective provided by the virtual camera. This is possible because, not only can the spatial configuration of an object be described in the terms of analytic geometry, so too can the modification and movement of such objects through space.

Because they are thus usually articulated in a way that allows for the rearrangement of their parts, polygonal models can be posed and animated much as physical maquettes or puppets might be animated by capturing sequential frames. This might involve "key frame animation," where the start and end points of the movement are defined, and the computer handles the transition between these; or the motion capture of an actual performer to capture especially lifelike movements of characters. More complex modes of animation such as procedurally generated movement, or the use of a simulated physics program to affect the objects within a given spatial environment, are possible. Simulated physics has played a role in computer graphics for a long time now, and its sophistication has led to impressive results. The Unreal graphics engine now has a "physics and destruction" system—appropriately named *Chaos*—that can depict real time environmental destruction of an almost cinematic quality. Not only can objects and their spatial environments be depicted, but the destruction of these things too. When CGI is combined with simulation, the *dynamization* of pictorial space imagined by Panofsky seems to hold the promise of being fully consummated.

All of this detail of how CGI allows for the pictorial impression of space is consistent with the account of spatial depiction presented earlier in this chapter, and fundamentally, computer graphics can be understood as a way of producing artefacts consisting of "colors distributed on a plane" that give the impression of spatial configurations of objects and scenes (Hyman, 2006: xvii). And, as I will argue, it is this pictorial medium that is a crucial part of VR's remediation of space. The images that are rendered on the screens of stereoscopic headsets are produced in this way. The polygonal objects and their animation, the placement of light sources, the role of the virtual camera, can all be found in the pictorial media of VR. And these techniques generate much of the sense of space in VR: the movement of characters and objects through an environment, lighted by different apparent sources, and captured by a virtual camera with changing focal lengths and framing, may naturally give rise to a sense of pictorial spatiality *as if* a scene were being filmed by an actual camera. However, VR adds extra elements to this basic picture.

3.3 Tracked Perspective

A further fundamental reason for the spatial dynamism in CGI is that the virtual camera itself can change its perspective to be moved *through* the graphical environment. In CGI, the mathematically defined position and orientation of the virtual camera need not be fixed, because the effects of its "movement" through graphical space can be defined and displayed on a 2D pictorial surface via coded mathematical transformations of the polygonal objects viewed. The movement of the virtual camera is of fundamental importance to 3D graphics, because for the three-dimensionality of the polygonal models to be fully revealed, the viewpoint must change for their hidden or occluded facets to be shown. Without the movement of the virtual camera, such models would remain little more than apparent graphical "sprites," an earlier (but still common) means of CGI depiction where objects are rendered as simple 2D bitmaps.

Indeed, this apparent movement of the virtual camera is key to revealing the full spatiality of CGI imagery and thus of generating the impression of spatial experience. Like the movement of an actual camera, the movement of a virtual camera through a scene can augment the sense of spatiality of the depicted scene, as the occlusion, relative position, and relative sizing of the objects and figures in the space changes to reflect the movement of the camera. Moreover, given that the camera can move *into* the graphical environment, the spatiality of this environment is revealed as the user explores more and more of the space. Compared to the static pictorial spaces we find in paintings and sketches, this is a radical development, but when compared to the art of the moving image it is not so radical. In the early history of cinema too, it quickly became evident that the camera could be moved into the environment to reveal more of the space and action of the scene. This visual technique sees its culmination in the long tracking shot, employed famously by Martin Scorsese in the film *Goodfellas*. In such cases, the space is revealed over time, and so the pictorial space gathers to it a temporal dimension.

The control of this movement by the user is a key aspect of many instances of CGI, and particularly videogames. For example, in *Minecraft* on the PS4, moving the analogue stick forward moves the camera "into" the spatial environment, and in the conceit of the game this amounts to the player "exploring the world." The player of the game can also "pan" the virtual camera side to side and up and down by simply moving the analogue control stick, and this is associated with the character in the gameworld moving their head. Literally, the physical movements of the player have the effect of providing an input that, via the algorithms in the game's program, generate a changed pictorial configuration that gives the impression of the changing orientation of the viewer in respect to the pictured scenes. But in practice, these pictures allow a user to "move into," "look around" or "explore," the environment depicted. I will have more to say on what this "exploration" really (or fictionally) amounts to in Chapter Five.

VR extends on the depictive technique of the mobile user-controlled virtual camera with the introduction of its second characteristic feature of *tracked perspective*. In tracked perspective the placement and movement of the virtual camera in pictorial space, both in its position and orientation, is made to *track* the user's own physical placement, perspective, and movement in their actual environment. As a simple example of this, in *virtual reality Minecraft* on PS4 VR, unlike its 2D screen antecedent, the player looks around the environment, not by manipulating a controller in their hands, but by simply turning their head to "look around" the actual space they inhabit. The player may thus look over their shoulder or even lean to look around the corner of a building by performing the bodily gestures comprising these actions in the actual world.

The very basics of tracking, introduced in the first chapter of this book, need to be explained in more depth, and placed in the context of the theory developed here that VR is a kind of pictorial medium. In PlayStation VR, this tracking technology involves small LEDs fixed to the exterior of the headset which are tracked across the image plane by a small camera set in front of the user (usually on the top of the television screen, which itself displays a 2D visualization or "Social screen" that those not wearing a headset can also view). The headset also contains an accelerometer and a gyroscope, and combining these with the tracking of the attached LEDs, the position of the headset can be tracked through six degrees of freedom—backward/forward, left/right, up/down, roll, pitch, and yaw—in physical space. The orientation of the user's head movements is then depicted by the apparent spatial orientation of the virtual camera in the depictive space by systematically altering the shape, relative position and sizing, occlusion, and so on, of the 3D models that constitute the graphical space. These, as we will find shortly, are then rendered for viewing on the stereoscopic headset to give the viewer the impression of an experiential space congruent with their own actual movements. The congruence of the user's movements and the apparent movement depicted in the graphical space is a central part of the pictorial illusion fostered by VR.

It is not just the user's perspective that is tracked, as the camera in the PS4 may also register and trace the movement of the controller in the player's hands, either in the form of the standard DualShock 4 wireless controller, or dedicated wands called the PlayStation 4 Move motion controller. Like the headset, LEDs and physical sensors allow for the physical orientation and movement of the controllers to be tracked through space. The PlayStation Move controllers enable each of the player's hands to be independently tracked, so that more complex interactions with the VR world become possible. Each wand has a small globe, lit by an LED, that can be traced across the image plane, allowing for 2D tracking. But because the apparent size of the globe can also be registered, the distance of the globe from the camera can be calculated, so allowing for it to be tracked in three dimensions. Each controller also includes a sophisticated collection of accelerometers, inertial sensors, angular rate sensors, and even a magnetometer, to allow the controller's precise orientation and acceleration to be traced in actual space, so that these can then be represented in depictive space, often associated with the movement of the player's virtual hands, or instruments within those hands.

Other VR systems allow the user's full bodily movements, including limited ambulatory movement around a room, to be tracked and depicted in the configuration of representational space. The HTC Vive VR system involves "room-scale" tracking, that is, the user's movements around a small room can be tracked into pictorial space, permitting the depiction of activities such as walking and turning around, bending down, and other physical gestures and activities. Technically, this is achieved by the Vive's use of two infrared tracking devices—"Lighthouse" tracking bases—that are placed in the corners of the room (which, of course may simply be an open space) and which track the position, orientation and movement of the headset and handheld controllers. Additional trackers can be added to objects or the user's body to provide further points of spatial data. Again, the movements of the tracked points in the room's space are then "placed" in the VR environment in a way that, when depicted in the stereoscopic headset of the Vive system, they give the appearance of an agent's movement within a graphical environment.

Perhaps one of the most interesting developments in VR technology is eye-tracking, such as that created by the VR technology developer tobii. The principles behind eye-tracking have long been known (Clay, König and Koenig, 2019), however, the technical ability to track the position and movement of the eyes, and have this be incorporated into pictorial techniques, has had to wait until the development of VR. Eye-tracking involves measuring the position and movement of the eye, allowing for the tracing of the focal point and the object of focus, and opens the potential for the depiction of the sensory effects of eye movement, including the effects of focus on the depth of the visual field. Of particular interest is the potential in eye-tracking technology for "foveated rendering," that is, where a VR depiction might render in detail only that portion of the visual field on which the user is focusing (Patney, et al., 2016). The technique is an opportunity to maximize the performance of VR headsets without a loss in perceived detail of the graphical scene because the *skimping* in detail occurs outside of visual awareness.

There are then several ways in which the movements of the user and the orientation of their head, eyes, hands, and body can be tracked, and for this tracking to be incorporated into the VR depiction. But what is the precise sense of the concept of *tracking* being used here, and how does this relate to the analysis of virtuality conducted earlier? Tracking is a two-part relationship: x tracks y, where, given some property of y, deviations or changes in that property are registered in x. So, for example, the changing position of a satellite in its orbit around the earth is tracked when these changes are registered in some form by a ground station. Moreover, the relationship is counterfactual in form: if the satellite had suddenly plunged to earth last Sunday evening, the ground station would have registered this event. The precise form of tracking in this instance is in terms of spatial information, and so the behavior of the satellite and its informational registration have a formal correspondence. If we think now of x as tracking more than one property of y, the more properties of y that are tracked by x, the more structural correspondence there will be between x and y. Of course, a great many features of x and y, not relevant to the tracking relationship, will differ, but an extensive and multidimensional tracking relationship between x and y may begin to embody a deep formal correspondence between the items.

This sounds a great deal like the relationship of isomorphism that I claimed accounted for the shared causal or functional efficiency between virtual and actual items of a kind. This explains the contribution to virtual reality media of spatial tracking devices. The movement of the camera in virtual depictive space is isomorphic with the movement of the head of the wearer of the PS4 VR headset because they can both be described by the same mathematical formulae in analytical geometry. The six degrees of freedom—left/right, up/down, backwards/forwards, pitch, roll and yaw—can be described in Cartesian space in a way that satisfactorily describes both the movement of the headset and the apparent movement of the virtual camera in graphical space. Because of this formal correspondence, the user may receive the impression of virtually looking around the graphical space surrounding them.

It might be responded to this, that, as described above, the use of a videogame controller to affect the depictive space is also a case of tracking, but not a case of virtuality, because the thumb and finger movements do not seem to correspond in an *appropriate way* to what is represented in the virtual world. Turning your head to look at an item in virtual space seems very different to manipulating analogue sticks to do so, and the former might seem a kind of fully virtual movement, while the latter is not. What this shows is that tracking by itself is not sufficient to establish that an actual movement translates into a virtual movement. There must also be a correspondence between the *kind* of movements involved in the tracking relationship. The virtual embodiment of shifting one's gaze, looking at an object, turning one's head—what virtual reality theorist Mel Slater refers to as "sensorimotor contingencies" (2009: 3550)—requires that it is these visual gestures themselves that are tracked and depicted in virtual space. This, essentially, is the difference between *gestural* and *non-gestural* control of a virtual application, a control variation that is often evident in VR video-games (Tavinor, 2018: 158). In fact, there are cases where controller inputs of the user *are* virtually incorporated into VR representations and do count as gestural controls. One example is haptic VR gloves, which we will encounter later. In this case, the control movements of the user's hands are isomorphic with the representational substance of the depiction, and so are virtually tracked in a way that also bears out a correspondence in gestural form.

3.4 Stereoscopic Viewing

CGI graphics of the kind discussed in this chapter for the most part is used as a means of producing 2D picture planes, even ones that are dynamic and which can be interacted with as in most videogames. Most computer graphical applications employ some form of 2D surface, and most videogames, of course, are depicted on a TV screen or computer monitor. This need not be considered a limitation in the pictorial form in such media, because these 2D representations can give rise to a real sense of spatial involvement and immersion. But what these graphical techniques cannot do is give the user a sense of being *within* the depictive space. For this, the medium will need a way to place the user within a virtual experiential space, and the most common way of achieving this is via a stereoscopic headset. Thus, to fully explain the notion of tracked perspective inherent in VR media, we need to discuss in detail how a visual orientation arises in native stereoscopic viewing, and how this can be tracked into the depictive space of a VR medium.

Native stereoscopic vision is a feature of human visual perception that allows us to perceive the three-dimensionality of our environment, including the location and solidity of the objects that furnish those spaces. Humans share stereoscopic vision with other animals, and it likely evolved because it allowed such animals to secure various adaptive goals such as locomotion and feeding behaviors (Heesy, 2009). The eyes of the average adult human are around six centimeters apart and this lateral displacement means both that the eyes will converge on external objects from slightly different locations, and that the images produced in those eyes by the light rays travelling from external objects seen will have subtly different shapes when registered on the retina. A consequence of this is that the three-dimensionality of the resulting visual experience comes about by at least two kinds of physiological information processing. First, as the eyes converge on the object, the signals from the extraocular muscles that control eye movement are registered by the brain and used to locate the position in space on which the eyes converge. This, naturally enough, is known as "convergence," and involves the visual system making a simple trigonometric calculation to triangulate the position of the external object.

Second, the geometrical differences in the retinal images allow for depth perception via stereopsis. The differences between the images registered by the eyes are called "binocular or retinal disparities" and are crucial to stereopsis because of the information they carry about the spatial distribution of external objects (Collewijn and Erkelens, 1990). The discrepancies are processed in the visual cortex and result in the perception of the positioning of objects in perceived space. Typically, stereopsis involves the "singleness of vision," that is, the retinal images being fused together into a single object, but this is not always the case as stereoscopic spatial perception likely comes in course and fine-grained forms (Wilcox and Allison, 2009). Fine grained stereopsis, which involves the retinal disparities being fused together into a single percept, allows for the fine motor control in actions such as threading a needle. Coarse grained stereopsis does not necessitate the fusing of the disparate images, meaning that the percept remains "diplopic." This may be key to our more general sense of the spatiality of our environment, particularly as it is viewed through peripheral vision (Wilcox and Allison, 2009). Stereopsis may also involve the differential monocular occlusion of *background* objects inherent in the retinal disparity, which is sometimes called "Da Vinci stereopsis" (Heesy, 2009: 30). In this case, the perceived object will occlude a different portion of the background for each eye, and the brain interprets this differential occlusion as a solid object, allowing it to "pop out" against the background. Stereopsis is likely the most effective when it involves perceiving and interacting with objects in near three-dimensional space, for example, in activities such as one-handed ball catching when moving objects need to be intercepted (Mazyn, et al., 2004).

In fact, much of our sense of visual three-dimensionality can be attributed to the vision of a single eye, utilizing effects including accommodation, monocular occlusion, motion parallax, and the relative and familiar sizing of known objects, some of which we encountered in the earlier discussion of the medium of moving images. Of particular importance in the later discussion here will be visual *accommodation*. Accommodation is a visual reflex action (often acting in concert with visual convergence, and indeed "coupled" to it) which involves the eye changing in focal power to fix on objects in the visual field. In humans, the focal length is altered by contraction or relaxation of the ciliary muscles, leading to an alteration in the shape of the eye lens. Information from these physiological events is incorporated by the visual system, providing information—though somewhat weak information—about the spatial location of objects close to the viewer. All these visuospatial cues (and others not detailed here) are employed by the visual system to perceive the depth and solidity of external objects, allowing for the interaction with these.

It is against this background that we can understand the varied means by which VR media allow for spatial visual experiences. Stereoscopic headsets are designed to function within this native visual context, though, as we will find later, their functional match with native vision is sometimes less than ideal. Drawing on the discussion of the previous sections, the basic pictorial content to be displayed by VR media comprises a geometrical model as seen through the tracked perspective of the virtual camera. This produces a 2D pictorial configuration that can then be rendered on a depictive surface. In most CGI animations this abstract geometrical configuration is rendered as a single image on a 2D screen such as a movie or television screen, or computer monitor. However, in the case of VR this situation is altered significantly in that the graphical model is configured for pictorial rendering by *two* virtual perspectives. This dual pictorial configuration is algorithmically produced to give the impression of two slightly offset perspectives corresponding to the actually offset images of the stereoscopic headset viewer. In the stereoscopic viewer, these two images are placed side by side—in the

PlayStation 4 VR headset, on a single 5.7-inch organic light-emitting diode panel with 1080p display resolution and an advertised potential refresh rate of 120 frames per second. The resolution and refresh rate—which contribute to the smoothness of the displayed movements—are important if the depicted world and its objects are to give the impression of solidity and permanence. Each image has a lens before it that widens the apparent field of view of the image, distances the focal point to a more comfortable position for the eyes, and softens the image to reduce the appearance of the so-called "screen door effect" which is caused by the visual prominence of the pixilation of the screen.

Each of these subtly different images is thus rendered on the screen of the stereoscopic headset, and when encountered by the viewer, the various aspects of the native visual system that allow for spatial perception described above, are engaged. Imagine that the VR system presents a virtual still life, comprising an apple sitting on a table in a room "before" the viewer. First, of course, the eyes *accommodate to the screen*, bringing it into sharp focus. However, the eyes also converge to the apparent position in the virtual depiction that the apple occupies. The convergence allows the physiological system underlying native vision to then judge the apparent distance of the apple. At the same time, by employing stereopsis and the retinal discrepancy of the dual images of the apple, the slightly offset images are fused, and so reconciled by the visual system as images of the same object seen from slightly different angles. The object thus fused might then be conceptually identified as an apple (or not, as the case may be).[4] Furthermore, the manner in which the offset images differentially occlude elements of the background can also be used to further judge the position the apple occupies and to give it a sense of solidity, so engaging Da Vinci stereopsis. Equally, other aspects of spatial perception—motion parallax, occlusion, relative sizing and so on—may also feed into this spatial perception, and they do so, because like convergence and stereopsis, the stereoscopic headset engages our native visual systems in much the same way as the natural world does. As a result, the visual system gives a profound impression that a solid object—an apple, by appearance—sits in a defined position before the viewer, perhaps tempting the user to reach out and pick it up!

Placing stereoscopy as a central feature here raises an interesting question: are stereoscopic viewers *necessary* for VR spatial experience? If monocular viewing can give a visual impression of spatial depth, then perhaps stereoscopy is just one means of depicting virtual space. An alternative to stereoscopic picturing has already been mentioned as a potential pictorial means of remediating spatial experience. The use of wrap-around parabolic projection screens by theme park "dark rides" can give an extraordinary sense of spatial experience. These are not stereoscopic viewers, rather they give a sense of spatiality because of the enveloping visual field, and how the shift of the viewer's apparent perspective in that visual field changes with their actual bodily movement. In *Harry Potter and the Forbidden Journey*, which I was lucky enough to experience at Universal Studios in Osaka Japan, the riders are buckled into a free hanging seat perched upon a robotic arm that is propelled along a track. Though much of the ride involves animatronic figures and practical effects, periodically the riders are encompassed with a hemispherical screen on which is projected from the rear of the screen a CGI animation of the viewer swooping and diving around Hogwarts Castle. The ride gives an extraordinary sense of movement through space, and perhaps is close to a kind of virtual medium, even though it lacks tracking and interaction.[5]

Another interesting possibility would be if the animated scene were rendered not on the surface of a stereoscopic headset, but on a frame on a wall akin to a painting, with the viewer's visual orientation tracked to produce changes in this surface dependent on their changing physical location before the screen. We might call this technique "Alberti's virtual window" because it would produce a 2D picture plane, though one, that because of the tracking of the viewer's perspective and the potential for the CGI animation of the scene, would allow for a dynamic pictorial space. In this case, the spatiality of the viewing would not arise from stereopsis, but from features such as motion parallax, occlusion, and relative and familiar sizing of the objects depicted within the pictorial space. This should still be considered a case of VR media, but one in which the picture appears as a portal into a space beyond the picture surface. This would amount to a kind of *virtual trompe l'oeil*, perhaps appearing much like Samuel van Hoogstraten's *A View Through a House*, though dynamic in viewpoint. In *trompe l'oeil*, a technique that often draws on the principles of linear perspective, the picture may give the (sometimes fleeting) appearance of an experienced space, but this is very much dependent on the orientation one has on the depiction. In the speculative Alberti's window, the requirement for the fixed viewing position could be solved, and as a result the window might give a stronger and more persistent impression of spatiality.

Hence, there may be VR methods besides stereoscopic headsets to generate the spatial environments we associate with VR media. The method of generating the spatiality of the depiction seems different in each alternative. In the virtual *trompe l'oeil* of Alberti's virtual window, the tracked actual position of the viewer is used to modify the apparent perspective of the picture plane. In the parabolic viewers, the picture plane encompasses the viewer fully, and its perspective changes to reflect the user's bodily representation of their actual spatial movement. This seems to show that in VR, like the depiction of spatiality in traditional painting, there is more than one way to cultivate in viewers the impression of space in a virtual environment, and as already hinted in the case of dark rides such as *Harry Potter and the Forbidden Journey*, not all these techniques need be visual ones. Let us now move on to inspect these alternative perceptual modes.

3.5 Sound and Body

We tend to primarily consider VR in terms of stereoscopic headsets and as a visual medium, but as characterized in the previous chapter, this is not the case. If VR media aim for the remediation of spatial experience, there are clearly other ways in which VR artefacts, apps, and games attempt to do so by non-visual means. VR is a *modal* representational medium, which includes visual depictions, but also sound, haptic displays, and even the representation of proprioceptive elements.

First and most obvious is sound, or "sonic virtuality," a term Tom Garner uses in his impressive analysis of virtual reality sound (2018). The PlayStation 4 VR system comes supplied with ordinary earbuds that plug into the system and provide a stereo image of the sound of the environment represented (and the other non-environmental elements of the application, such as a musical soundtrack). Stereo sound has long given listeners a sense of spatiality in other media: so, for example, the multitracking of audio recordings that was developed primarily in the 1960s allowed artists such as The Beatles and Pink Floyd to create complex spatial sound recordings where audio sources might appear to originate from the left or right of the listener or pan from one side of the space to the other. Audiophiles long valued Pink Floyd's *The Dark Side of the Moon* because of the separation in its stereo sounds and how they were used to create various effects. And in cinema, of course, stereo sound also has a long history. Anyone who has watched one of Christopher Nolan's recent films in a cinema will know how modern sound production can lead to surprising effects. (And how loud it can be.)

However, the fully fledged virtual 3D audio I am interested in here amounts to something more: it allows for sound sources to be spatially *localized* for the listener, so that sound sources can be placed in specific locations in virtual space. Unlike a stereo headset where the apparent sound sources are indexed to the headset, in virtual sound, when the user turns their head, a virtual sound source will give the impression of being stationary in 3D space. Again, this is a remediation of spatial experience, modelling in a virtual way a native perceptual process. Spatial audition is complex and involves multiple sound cues and has been poorly understood until relatively recently (van der Heijden, et al., 2019). The basis of spatial hearing is the discrimination of the waveforms received by the two ears from sound sources. Because the ears are separated in space, a sound arriving from a source close to left of an individual will hit the left ear milliseconds before the right ear; and the right ear is also in the "sound shadow" of the head. The resulting "interaural" differences in the waveforms are processed by the brain allowing for the perceived localization of the sound source. Various technological methods have been used to remediate spatial audition, but "binaural rendering"—the combination of headphones or speakers, and spatial tracking—is perhaps the "most convenient and accurate technique" because of its simplicity (Nguyen, et al., 2009).

The music technology manufacturer Boss has produced a virtual guitar amplifier, the "Waza Air Personal Guitar Amplification System," that gives a good example of binaural rendering, and that should be considered as an example of virtual media in virtue of this. Portable headphone guitar amplifiers have been common for a long while and they might give a sense of stereo sound. But what the Waza Air achieves is to locate the apparent sound source of the modelled guitar amplifier in a virtual spatial position relative to the player. The headset includes a gyroscope to allow for this spatial tracking of the guitar player's movements and the resulting audio space that the player is "in" to be modelled: one setting has the amplifier position behind the player as though they were performing on stage with a band. Other proprietary 3D audio systems such as Dolby Atmos, Apple's spatial audio for its AirPods, and PlayStation 5 3D audio are all new applications of this VR concept. PS5 includes a 3D audio engine called Tempest, that is an audio development of the VR technology debuted on the PS4, and it uses a graphics chip to process the "object-based" sound sources for sound localization and binaural rendering.

Haptic technology involves the conveyance of information through the medium of touch. Haptic displays have long been a factor in videogames and now also play a significant and expanding role in VR media. In videogames the most common form of haptic display is vibrating controllers such as PlayStation 5's "DualSense" haptic controller, or "force-feedback" where a joystick or racing wheel might move to depict the forces acting on the controller in the world of the game. Haptic controllers typically provide quite non-specific feedback about the tactile or physical forces being encountered in the virtual environment, though the controller for the PS5 has recently extended its functionality with "adaptive triggers" that adjust the feedback depending on the corresponding action the trigger represents in a given game.

More ideal is to have a peripheral such as a glove or vest, that fits on some part of the body to both allow for the tracking of the body part and provide sensations corresponding to virtual world events. Haptic gloves are the most common of such devices and have recently undergone rapid development. Commercial developers such as *HaptX, VR glove,* and *Teslasuit* are all producing haptic gloves and other haptic peripherals for use with current VR systems. It is worthwhile discussing just one of these to see the functional potential of the device, and how it fits into the account of VR developed here. "Dexmo" haptic gloves, produced by Dextarobotics, a company based in Shenzhen China, simulate the experience of grasping objects by providing force-feedback to the individual fingers. The glove is tracked in VR space, and as the user grasps a VR mediated object, servos on the glove restrict the finger movement to give the tactile impression that one is grasping a solid object. Again, in terms of the general theory of VR offered here, this haptic device remediates an experience we might have in actual space, that is, the tactile impression of the solidity of objects. The potential uses of such a VR device are obvious, extending from gaming, to education, and one suspects, the military. Furthermore, we can easily see that such gloves count as a kind of gestural control referred to earlier.

Finally, kinesthetic displays are also possible, and these typically provide a virtual remediation of proprioceptive experience of space and movement. There is a little bit of ambiguity between haptic and kinesthetic feedback, and technically, some of the haptic devices above might be properly considered kinesthetic devices or may have kinesthetic in addition to tactile aspects. Nevertheless, it is more common to associate kinesthetic feedback with those devices that provide a sense of bodily movement. Proprioception, or kinesthesia, comprises the ability of the body to sense its own special position and movement. Mechanosensory neurons exist in many parts of the body such as muscles and joints and provide information to the brain pertaining to various kinds of bodily positions and motions. These neurons work in concert with the vestibular system and vision to allow the brain to track the positioning, movement, and acceleration of the body though space. The obvious way of achieving a sense of bodily movement in VR is by actually moving the body of the VR user. This technique has a relatively long history outside of VR, including theme park rides such as Disneyland's *Star Tours* (1987) which included a hydraulically controlled base, allowing the ride to be tilted through six degrees of freedom, simulating in-flight forces; and any number of flight simulators that virtualize physical forces by similar means.

Such systems are also used in conjunction with VR headsets and visual displays. One increasingly prevalent example of kinesthetic displays are the robotic arms used in dark rides such as the earlier mentioned *Harry Potter and the Forbidden Journey*. These arms, often attached to a track, can give riders the sensation of moving through space by moving and tilting the body of the user to give them the proprioceptive sense of acceleration. To simulate forward acceleration, the arm tilts the user to lie flat facing upwards, so that the gravitational forces seem to the user—usually when combined with animations of a wrap-around parabolic projection screen—as the forces of forward acceleration. The apparent movement produced by these forces tracks the movement depicted on the screen. Robocaster is a manufacturer of these arms and has supplied them to many theme park rides. These commercial rides allow for the virtual tracking of movement in a spatial scale that exceeds the limited scale of haptic force-feedback available in personal VR devices.

This is far from an exhaustive treatment of the kind of displays now possible in VR; and with the rapid advances in the medium, it is necessarily incomplete. Nevertheless, we have found in this chapter that there are many components of VR media, and they are not restricted to a visual modality. A ball might be depicted via a visual configuration on a stereoscopic headset; the sound of it bouncing on the ground next to the user might be placed in virtual space by localized audio; and even, if it were able to be virtually picked up, its tactical representation is possible through haptic gloves. We can consider these components separately, and each has a separate technological lineage to be traced and explained, but all these virtual media work in concert to produce the ultimate spatial effect of VR. Moreover, in their combination, we see something significant about VR media: that *it is a collection* of changing and expanding technologies, and not one unitary "thing." But we can reiterate on why this collection of individual technologies counts as a virtual medium, that is, what binds these media together. These media, employing geometrical spatial models, spatial tracking, a stereoscopic headset, haptic gloves and other perceptual modes, aim for the functional efficiency of a real perceptual engagement in a spatial environment, though in a technologically and non-customary way. Again, of course, it is still an open question whether VR media do successfully remediate spatial experience: just as VR seems to embody the aspirations of earlier artists in this regard, it may also repeat the mistakes of that approach.

Notes

1 Readers looking for more technical substance on the technology of VR might like to consult Steve LaValle's very detailed online lecture notes on this topic (2019).
2 See for example, Walton, 1984, and Cohen and Meskin, 2004.
3 One potential complication here that I do not have the space to discuss, is that even the "real" scene might be "fake" in various respects, by involving building facades, matte painted scenes, or practical effects such as forced perspective. So, even in the case of the filming of live scenes, artifice can be used to give a false or contrived impression of spatiality.
4 I am leaving aside discussion of how the recognition of the object of visual perception occurs until the next chapter, where it will be relevant to understanding how VR functions as a pictorial medium.
5 A related case is 360-degree or immersive video, where an omnidirectional camera is used to film in all directions at once, so that the vantage point of the user can be changed during the viewing of the image. This technique is also sometimes referred to as VR and is sometimes adapted for use with 3D viewers such as Google Cardboard and Samsung Gear VR. Nevertheless, it seems a superficial form of VR because it lacks the tracking and interaction associated here with fully fledged VR.

4 VR as a Picturing Medium

4.1 VR Media and Picturing

In the previous chapter I framed the investigation of VR historically, in terms of previous attempts by artists to remediate the experience of space, particularly through the pictorial technique of linear perspective. VR has several technical advantages over the linear perspective associated with Alberti and Brunelleschi, including its animation of objects and spaces, the dynamism of the viewer's perspective on those spaces, the binocularity of their visual perspective, and indeed different perceptual modalities—hearing, touch, proprioception—which all add to the realistic spatial impression we associate with VR. Via the visual techniques alone, VR media make what seems to be a novel contribution to experiential remediation: in VR, the user takes an apparent position *within* depictive space, in that the virtual perspective constitutes a spatial anchor or index, or a pictorial "here." Thus, suddenly, the notion of something being pictorially to your left or to your right, or behind or before you makes sense. This differs profoundly from earlier cases of depiction, where the user's placement and perceptual orientation are not embodied in the spatial configuration of the depictive surface. What results, I suggest, is a kind of *egocentric picturing*.

The foregoing account of VR technology may result in a challenge to our notions about picturing, as egocentric pictures seem very different to those with which we are mostly familiar. That VR presents visual images that give the viewer an active perspective within the picture space, are dynamic, and that they seem to allow the user's interaction, might lead some to question whether they are pictures at all. The current chapter takes the technical discussion of the previous chapter and situates it in the philosophical perspective of the ongoing debate about the nature of picturing and picture perception. It addresses some of the philosophical issues with seeing VR as a pictorial form, but also argues that if it is to be considered as such, VR will require us to alter our conception of what pictures are, and the modes of experience and interaction they allow.

In his philosophical and psychological account of pictorial seeing, Bence Nanay observes that, regarding typical pictures such as paintings and photographs,

DOI: 10.4324/9781003107644-4

> [...] we don't and can't perform actions on depicted objects. Further, a
> minimal condition on performing perceptually guided actions on objects
> is representing the spatial location of this object in one's egocentric
> space. As in front of us, to our left, etc. If we couldn't represent the
> spatial location of an object in our egocentric space, then we would
> have no idea which direction to reach out to grab it or use it for any
> other action. But, crucially, depicted objects are not represented in our
> egocentric space: the depicted space is not our egocentric space.
>
> (2015: 189)

Here, Nanay makes two related and reasonable claims about pictures. First, the
depicted content of pictorial images is not the kind of thing that viewers can
interact with. Attempting to grasp at an apple depicted in a gallery painting
would not only be perceived as odd by onlookers, but it would also count as a
fundamental misunderstanding of the depictive capacities of paintings. Second,
and partly explaining this, pictures do not represent their objects in the egocentric space of the viewer. That is, when we look upon pictures in a photo
album or gallery, these images typically give no indication of where the items
depicted in the picture are placed in relation to the viewer. Viewers can of
course position themselves in relation to the *surface* of the image, and can frequently interact with and alter this surface, but the content depicted by that
surface is insensitive to both our presence and our activities.

The technical developments in virtual reality media described in the previous
chapter places these claims into question, at least if we regard VR media as a
form of picturing. In the PlayStation 4 VR game *The London Heist* (2016),
employing a stereoscopic headset and motion tracking technology, players can
locate the depicted objects in their egocentric space, and they can reach out,
grasp, and use many of the objects depicted in the game. In one sequence, the
player finds themselves in a smoky English pub in which they may reach out to,
and pick up, a cigarette lighter on the table before them, using it to light a cigar.
There are complications here, of course—the smoky English pub and the
cigarette lighter are fictional things—but that the picturing technology in this
case allows for the *appearance* that the player is related in egocentric space to a
pictured object with which they can interact, by itself provides a challenge to
Nanay's claims.

Nanay's account of picturing is a version of the "twofold" model of picture
seeing that holds that when we view a picture, we perceive at least two
things—the picture surface, and what is depicted by the surface—and that we
see these simultaneously (Nanay, 2018: 165). A classic statement of the twofold
theory owes to Richard Wollheim (1980; 1998). Key to Wollheim's theory

> Is a special perceptual skill, called "seeing-in," which we, and perhaps
> the members of some other species, possess. Seeing-in is prior, both
> logically and historically, to representation. Logically, in that we can see
> things in surfaces that neither are nor are taken by us to be

representations, say, a torso in a cloud, or a boy carrying a mysterious box in a stained, urban wall. And historically, in that doubtless our remote ancestors did such things before they thought of decorating the caves they lived in with images of the animals they hunted.

(Wollheim, 1998: 221)

"Seeing-in" is argued to explain the phenomenology of seeing a picture's surface and seeing what is represented in that surface, in that when we look "at a suitably marked surface, we are visually aware at once of the marked surface, and of something in front or behind something else," a feature that Wollheim calls "twofoldness" (221). Twofoldness refers to a single experience with "two aspects," namely "configurational" and "recognitional." The former is involved in seeing the configuration of the surface, and the latter in the recognition of the scene, objects, or figures depicted. At once, then, we may see the colors and shapes of the surface and the philosophers and apples they depict. Wollheim's theory is only one amongst several theories of pictorial seeing, including "semiotic" and "resemblance" theories. It has also been criticized (Walton, 2002; Levinson, 1998). It is outside the scope of this book to give the theory a thoroughgoing defense from its critics or evaluate it against alternative theories, though one frequent complaint about the theory is fair: that in Wollheim's formulation of the theory the precise nature of the perceptual act of "seeing-in" remains a little vague and that there is "little explanation" of precisely what twofoldness amounts to (Walton, 2002: 33).

It is against this background, and partly with the intention of remedying the explanative gap in Wollheim's theory, that Nanay approaches the topic of the twofoldness of pictorial seeing. One of Nanay's overriding intentions in his work on pictures is to provide a theory, satisfying both to philosophers and psychologists, of precisely what the simultaneity relationship between seeing of the picture surface (configuration) and the pictured scene (recognition) involves (2015: 185–193). His answer is that it is a matter of how two visual subsystems interact to construct 3D scenes out of markers on 2D surfaces (Nanay, 2008). Picture perception comprises the interaction of the "ventral stream" which is responsible for object "identification and recognition," and the "dorsal stream" which is responsible for the "visual control of our motor actions" (2008: 975). In picture perception the former allows us to perceive just which objects are depicted in the picture, while the latter allows us to perceive the marked picture surface in our egocentric space. The way in which these two streams interact explains features of picture perception, such as why an oblique viewing angle need not distort the appearance of the objects within a picture, because of how the ventral stream may adjust to information drawn from the dorsal stream (977).

There is a great deal more to Nanay's account, and a lot of evidence is mustered in its defense, but my focus here is how as a theory of picture perception it may be at odds with VR visual media. The difficulties that VR poses for Nanay's account of picturing stem directly from this twofold account: it is because the dorsal (configurational) stream cannot attend to the pictorial scene that "depicted objects are not represented in our egocentric space"; only the surface is thus depicted. But VR pictures plausibly *do* locate the depicted objects in the viewer's egocentric space; so much so that users often get a strong impulse to reach out to those depicted objects and may experience an often quite overwhelming feeling of "presence" within the depicted space (Witmer, and Singer, 1998).

We have at least three options of how to respond to this problem: we could, first, reject the idea that VR involves picturing and claim instead that it comprises some other representational mode that allows for egocentric viewing. Second, we might conclude that while VR is a mode of picturing, it does not allow for egocentric seeing or interaction, but only appears to do so. And third, we might decide that theories of picturing such as Nanay's need to be revised to accommodate virtual reality media as a technologically novel case of picturing, and so deny at least some of his claims about the psychology of picture perception. In this chapter I am going to defend the final option, partly because twofold theories of picturing in other ways do such a good job of explaining many of the technical and perceptual features of virtual reality media seen elsewhere in this book. VR media can be reconciled with the twofold theory of picturing, but the twofold theory may not come away from this reconciliation unscathed.

4.2 VR, Models and Simulations

So, by way of assessing the first option, why might we be tempted to conclude that VR is not a picturing technology? A first possibility is that VR is not a picturing medium at all, and that there is another way to characterize VR media that does not lead to the problems that egocentric or interactive picturing present. We might, to take two alternative conceptualizations, treat VR media as a means of *modelling* rather than picturing subjects, or as comprising a kind of *simulation*.

The games theorist Rune Klevjer is one to deny that VR representations can be adequately explained as pictures and in terms of theories of picturing, thinking them instead to be more akin to models such as dolls (Klevjer, 2017). Klevjer's set up of the problem is insightful, especially when he notes that because of the way we perceive and interact with them,

> games appear to be comparable to Lego sets, architectural models, or theme parks, none of which we tend to think of as pictorial phenomena. On the other hand, the modelled environment of video games, unlike such phenomena, is only accessible to us as images projected on a screen. How to conceptualise this duality has implications for how we think about the relationship between games and other visual media, especially cinema, and it also informs how we understand the role and nature of depictive representation in contemporary digital culture more widely.

> (Klevjer, 2017: 1–2)

Klevjer is right to pick out the tension here: games are interactive, and they do seem, in some ways, like toys or models. *Cities: Skylines* (2015), a city-building videogame, is a kind of virtual toy that one interacts and plays with, even as it is revealed through images on a screen. Klevjer goes on to deny that such VR models are depictions, claiming instead that they are more like physical dolls. This is because he denies that the content of pictures can bear a spatial relationship to its viewers, whereas dolls,

> Because they represent themselves as physical objects in space [...] are ontologically different from images, which appear in perception in a very different way, projected on a surface.
>
> (2017: 6)

In his comparison of VR with physical models, Klevjer also denies that pictures can be interactive: "So, unlike depictions, physical models invite actions that go beyond mere looking, either implicitly or explicitly" (Klevjer, 2017: 11) and while a VR model "would not be a physical model in the strict sense, ... it would have a similarly concrete, hands-on presence" (13). VR models, for Klevjer, are more akin to simulations than pictures, and he is tempted to take up what seems like a strongly metaphysical take on virtuality. Klevjer is likely motivated here by the same issues identified by Nanay, that is, that in pictures the object of appreciation is not in our egocentric space and cannot be interacted with, whereas in the case of virtuality this does seem the case.

In response, it should initially be noted that Klevjer has a rather inclusive notion of virtuality, using the term to refer to the 2D visualizations of computer games and video installations, in addition to VR media as I have characterized them here. Because of this, Klevjer's examples of virtual objects that exist in our physical space to be interacted with are less than convincing. He describes a piece of video art, *Notio Viri Placet* by the Norwegian artist Bård Ask, where the interaction of individually depicted singers gives the impression of a unified performance. The singers were filmed looking at the camera, and Klevjer argues that "the look of each individual singer, into the camera, unavoidably situates our view in relation to the space of the image ... as if they were present and functional in space" (12). However, despite its content, this example is a fairly ordinary non-egocentric video display, and though the singers may seem to be "following us with their eyes," this is a standard feature of many kinds of picture. In fact, that in a picture the eyes of a depicted person appear to follow you *wherever* you stand before the picture—as would seem to be the case in *Notio Viri Placet*—is good evidence that the pictorial content is insensitive to the viewer's spatial position, and hence that the picture does not "situate our view in relation to the space of the image."

But even if Klevjer had provided better examples, which are available, I will argue here that pictorial content can have the apparent spatial and interactive qualities he contends only models to have. Klevjer abandons picturing too quickly, and when he concludes that "Husserl and traditional theories of depiction can hardly help us here" (9) he is premature because he has not rigorously explored the potential that traditional theories of depiction have in this instance. Klevjer and I differ in that he thinks the genuine differences in virtual displays disqualify VR models from being pictures, whereas I see VR as a novel, though precedented depictive mode.

In some ways, taking VR representations to be models rather than pictures does account for a number of the features of VR, giving Klevjer's assertions some initial plausibility. It is a common locution to speak of computer representations as comprising models, particularly within CGI, where wireframe models are a key aspect of 3D animation. Unlike a picture of an airplane, a model of an airplane can be located in space, picked up, and manipulated because it exists as physical artefact. A wireframe model too, is a manipulable physical artefact, in some sense at least. Adding to this the prevalent notion that computer models can comprise *simulations*, we can recognize that, again differing from pictures, computer models allow for the structures and behavior of modelled objects not only to be represented on a screen, but to be predicted, observed, and measured. A computer model of an airplane can be placed in a model of physical space to simulate the forces to which an airplane is subject in flight.

Thus, a customary way to refer to the computer artefacts involved in virtual graphics is "3D model," and it is the terminology used in the previous chapter. But this way of speaking does not imply that such things are literally spatial models as Klevjer contends. In what sense are these mathematically defined artefacts three-dimensional or spatial *in any sense*? What we do not mean here is that the computer *model* exists in space as a three-dimensional object, because if it did the precise location of the space it occupies could be specified. The computer hardware on which a program is run exists materially in three dimensions, but a coded object such as a character model cannot coherently be thought to occupy a given three-dimensional space within that hardware, as though you could take the casing off a computer to point to the location of the modelled airplane within. Much less does the 3D object exist on a precise location of a screen, because the screen comprises a configured surface that is also quite unlike anything we might consider as a model of an airplane.

In fact, we are likely referring to two things when we refer to 3D models. First, the term is a way of referring to the computational artefact that codes the vectors, polygons, textures and so forth that are employed in the production of 3D graphical images. This ultimately does have a material instantiation in the circuits of a computer, but the code itself is abstract and nothing at all like a physical model. Secondly, the term may refer to the graphical artefact itself, that is, what is produced by these coded artefacts when they are rendered in some form. But here too we must be careful, because that these rendered artefacts are themselves three-dimensional is also ambiguous. Some such artefacts may be literally three-dimensional: this occurs when a 3D model (of the first sense) is employed in the process of 3D printing to produce a resin object such as a figurine. In this case, the polygonal model provides the dimensions, shapes and so forth that are used as a prototype to produce a physical artefact. But other rendered artefacts are not literally three-dimensional. 3D graphics can be employed to produce static 2D pictures nearly identical with the pictures produced by traditional modes of picture making. Many of the images we see in magazines or on advertisements these days are renders of 3D graphical models and these may well give an impression of spatiality via traditional pictorial means. The dynamic 3D images we are principally interested in here—those seen in videogames and VR media—are also typically rendered on 2D picture planes, and they too give the impression of being spatially configured. Hence, what 3D means is not that the computer artefact is literally spatially dimensional, but that it is apt to produce displays that give a sense of being configured in 3D space.

These ambiguities also exist in references to computer simulations. A simulation of an aircraft responding to the physical forces of flight is at its most fundamental a mathematical model that need not even be depicted to a viewer; but when it is depicted, its *apparent* spatial existence is a matter of its configuration on a picturing surface. The simulation differs to the basic graphical model, however, in that its computational artifact codes for much more than a basic physical appearance, including, perhaps, physical forces such as drag, lift, gravity and so on, that affect real aircraft in flight. These coded forces may affect what appears on the screen when the aircraft model is rendered as a graphical artifact, but this does not alter the "ontological" status of the picture thus produced but is simply a variation in how and for what purpose the picture was produced. This is a lesson I will later extend to VR pictures.

Thus, *pace* Klevjer, a virtual model is fundamentally unlike a physical model. Unlike an actual physical model, which may be immediately experienced and interacted with in an unmediated way because of its physical nature, computer models must be rendered to become the possible objects of experience and interaction. While it is frequently appropriate to consider VR representations as virtual models or simulations, when these things are visually presented to the user in a graphical way, they nevertheless involve picturing. To remove Klevjer's noted tension between this understanding, and the fact that virtual models can be manipulated and played with unlike most pictures, we need only explain how it is possible to manipulate or play with virtual objects, something to be considered in the final section of this chapter.

4.3 VR and the Picturing Surface

A second reason why we might doubt that VR is a kind of picturing is that there are concerns about the nature of the "surfaces" involved in VR, which seem very different to the pictorial surfaces found in other depictive media. That VR headsets typically display *two* subtly different images viewed through binocular lenses, may disrupt the idea that VR pictures involve a simple picturing surface. Second, there is the question of whether users of VR see the surface at all.

First then, does VR involve a picturing surface? And if so, what does that depictive surface consist of? The obvious place to locate the surface is in the pixel array of the stereoscopic headset. But this is a bit more complicated than it might initially seem. The PlayStation VR headset, as already noted, includes a small organic light-emitting diode panel that sits very close before the user's eyes. The screen is viewed through two lenses, which allow for a wider field of view and soften and magnify the images, disguising the pixelization of the array. The two images are algorithmically generated from a computer animated environmental model such that the apparent displacement of the images, combined with the binocularity of the lenses, mimics native visual stereopsis, and gives an impression of our natural visual situation in the world. VR clearly does not involve a simple surface such as a canvas or movie screen.

In response, I can point out that not all cases of VR have these technical complications: if hemispherical projection screens and what I have referred to as "Alberti's window" are cases of VR, then VR need not involve the complications of stereoscopy. But it is also not clear that even in the cases where VR does involve stereoscopy, that this disqualifies it from being a picturing medium, because there are already similar variations in non-controversial cases of picturing surfaces. Toys like the stereoscopic picture viewer View-Master, and 3D movies, both involve dual offset images and lenses that allow for stereopsis, and both are naturally classified as cases of picturing, and potentially can be understood through the twofold theory of picture seeing. Nevertheless, what we learn through the investigation of the surfaces of VR is that picturing may involve technological innovations that inaugurate new technical means of image production. Another striking example in which a compound pictorial surface gives the impression of 3D imagery is *holography*, in this case comprising a recorded "interference pattern" caused by different light sources striking and being reflected or scattered by an object.

Secondly, do users of VR see this screen? The answer seems obviously "yes," because the screen is right in front of the user's eyes and without this visual encounter VR imagery would be impossible: ultimately, it is this surface that encodes the visual information that allows for the perception of the scenes depicted in VR. Never in my experience of VR, was I not also *aware* that I was viewing images on a screen, and indeed that the features of this screen were perfectly evident if I attended to them. Admittedly, the designers of VR technology do spend a lot of effort in trying to disguise the screen, and both the so-called "screen door effect" caused by the pixelization of the screen, and the restricted field of vision inherent in stereoscopic headsets, are frequently considered faults in VR because of how they bring attention to the screen surface. As noted, one job of the lenses in a stereoscopic headset is to soften the screen door effect. But this does not mean that the designers strive to make the screen *invisible* so that it cannot be seen; rather, they seek to make non-depictive artefacts of the screen less visually prominent to the user so as not to inordinately distract from the depictive content. But even if users do not always notice or attend to the surface, they surely see it, because it is the surface that encodes the visual content that gives rise to the apparent spatial illusion associated with VR.

What is coming out in the discussion above, is that the notion of *seeing* is ambiguous between sensation, perception, and attention. In the most basic terms, sensation is the physiological process via which parts of the body—sensory organs—detect or register elements of the environment. In vision, rays of light travelling from objects in the external environment enter through the lens of the eye and cause chemical changes in the cells of the retina. The changes are transduced into "action potentials" which are carried into the central nervous system. Perception involves the selection, organization, and interpretation of such sensory information. Not only does perception involve sensory information, but it draws on memory, expectation, and attention. In vision, information of various forms, transduced from the eyes, and including the position, intensity, and color of light in the visual field, is selected and interpreted to build up a visual scene, including, importantly for our discussion here, a spatial interpretation of the environment being viewed. Finally, attention, while involved in the selection or amplification of perceptual features, can also refer to the *reportable* aspects of our conscious experience, that is, the parts of our perceived environment that are being consciously attended to (Chun and Wolfe, 2001). This is obviously a simplified analysis of these aspects of vision—and I apologize to psychologists and philosophers of perception who will no doubt find them oversimplified—but it is detailed enough to allow us to identify the ambiguity in the current concern.

Which is the relevant meaning of "see" for two-fold pictorial seeing? One must of course *perceive* the three-dimensional configuration pictured, but what is it to see the *surface* of a picture that supports this act of perception? While sensation is necessary for seeing—it is a *sine qua non* of any kind of vision—sensing a pictorial surface is not sufficient for pictorial seeing, because one may not recognize the pictorial surface as a picture, and not see anything in or through it. This might happen if the picture was viewed under less-than-ideal circumstances or was simply not recognized as a configured surface. This already shows that perception of the pictorial surface—the visual selection, organization, and interpretation of the surface features, and the building of this into a visual scene—is also necessary to seeing the surface of a picture. But what about the third aspect of seeing referred to above, that is, attention? Is this also necessary for pictorial seeing? It seems not. First, there are rare cases where even if viewers do see the marked surface of the picture, and perceive the scene depicted by that surface, they may mistake what they see for a real object or real environment, either again, because of less-than-ideal viewing conditions or because the surface has been expressly designed to give this illusory appearance, as in *trompe l'oeil*. This shows that the question of whether the surface of a picture is perceived is itself ambiguous between "perceiving the marked configuration of a surface," and "perceiving the pictorial surface *as* a pictorial surface." Second, even if the picture surface is recognized as such, one need not pay the surface of a picture any great attention, and still perceive the picture, perhaps when one is simply more consciously interested in the pictorial content than any features of how the surface marks out this content. Attention to the picture surface is not necessary for pictorial seeing.

It is thus perception, and specifically, the perception of the configurational aspects of the marked surface, that is thus the relevant interpretation of "see" in the question, "Does the user of VR see the screen?" And it is clear, given the discussion of how three-dimensional scenes are perceptually derived from the dual images of the stereoscopic headset, and the features of the visual encounter with these stereoscopic images to be discussed below, that users of VR do perceive the configuration of the pictorial surface. Moreover, users are usually somewhat aware of—if not paying particular attention to—the surface. The phenomenology of stereoscopic headset use, partly because of the infancy of the technology and its resulting perceptible artefacts, is ordinarily one of being aware of the screen.

But to be charitable to the contrary position, let us approach this question from the opposite direction: how might users perceive the contents of VR media *without* seeing the surface? There are at least three possibilities here. First, that users of virtual media experience the picture content from a position *within* the world depicted, might be thought to imply that VR involves seeing—perceiving—something like an encompassing visual field rather than a simple depictive surface. Second, it could be that users of VR are taken in by either a *trompe l'oeil* or an *illusion*, and while they sense the pictorial surface, in such cases they may not perceive it as such. Finally, some writers have suggested that when using VR, what the user sees—senses and perceives—is not the screen, but virtual objects themselves. This is a metaphysically strong interpretation that VR allows users to directly experience "virtual objects" or "alternative realities."

First, the suggestion may be that users do not see a screen but experience an "encompassing visual field" (Tavinor, 2019a). Pawel Grabarczyk and Marek Pokropski's recent account of VR immersion gives this impression, as they think a prerequisite of VR immersion "is to believe that there is some kind of alternative place in which we can immerse ourselves or be present (as opposed to a simple picture or an animation we can only look at from the outside)" (2016: 35). Experiencing VR media through a stereoscopic headset certainly does not feel much like viewing a picture "from the outside" because the depicted scene is all around you, and indeed one can typically move through it. Nevertheless, there is reason to question exactly how encompassing the visual field in VR is. The field of vision depicted in stereoscopic headsets is typically limited, with the PS4 headset allowing for a visual field of only 100 degrees. This limitation makes it perfectly evident in almost all cases of stereoscopic headset use that one is viewing a screen. But more importantly, to think that the object of seeing is an encompassing visual field rather than a depictive surface is merely to confuse the two folds of the picture experience in VR. The user sees (and perceives) the screen, but what they see *through* this perceptual encounter is an encompassing visual field, the latter being precisely the experiential egocentric space that I have argued to be a characteristic feature of VR picturing.

Second, it is sometimes held that VR aims for complete illusion and it is not clear under such circumstances whether users would see the surface in the way the twofold account requires. The idea that VR is a form of illusion is reasonably common in the literature on VR. For the art theorist Oliver Grau, the promise of VR media is the ability to "convey to the observer the illusion of being in a complex structured space of a natural world" (Grau, 2003: 15). Similarly, games theorist Aubrey Anable argues that "one of the primary goals of VR technologies is to do away with the screen, or at least the sense of the screen as a mediating interface, altogether" (2018: 47). This "place illusion" is purportedly so striking that it generates the previously noted phenomenon of "spatial presence" (Slater, 2009). It might be thought that if VR aims to produce compelling visual illusions, that for the user to see and be aware of the screen would be a *failure* of this intention, and so, the idea that VR media involve twofold picture perception is at odds with the nature of VR as illusionistic.

Granting for the moment this illusionist analysis of VR, we should ask whether a case in which the viewer experienced the "illusion of being in a complex structured space of a natural world" would involve seeing the pictorial surface. If seeing here is interpreted as "sensing," then the answer will be "yes" because the surface is what is visually encountered and is the proximate cause of the visual experiences. Moreover, the color plane of the surface is also perceived because it is the structure of the surface that is perceived as the spatial configuration which gives rise to the illusion. In fact, it does not seem to be an essential part of being taken in by a pictorial illusion that one need be unaware of the pictorial surface. It may be that being aware of the pictorial surface but continuing to perceive the illusory effect is what gives the pictorial illusion in VR its striking effect. This resistance to conscious awareness is a characteristic part of viewing other illusions: in the case of the Müller-Lyer illusion, a famous optical illusion first created by the German sociologist Franz Carl Müller-Lyer, when I look at it, I know that the horizontal lines are the same length, but I cannot help but see them as different.

Could it be that, in at least some cases, the illusion of VR is so compelling that the viewer is not aware of the surface? The best parallel here, perhaps, is with *trompe l'oeil* paintings, which we have already met on a number of occasions in this book. In *trompe l'oeil*—in French, "deceive the eye"—the artist produces a pictorial surface that, when viewed, gives the illusion of three-dimensionality. In some cases, the impression may be so strong that the viewer does not realize they are viewing a picture, because they are not aware of the pictorial surface. In a discussion of Wollheim's theory, Susan Feagin tells the entertaining story of a friend who misperceived Raphaelle Peale's 1822 painting *Venus Rising from the Sea*, where Venus is all but obscured by a *trompe l'oeil* napkin (Feagin, 1998: 237). I am a little ashamed to say I had this experience, very briefly, with a large (and rather inexpert) *trompe l'oeil* painting in the California Adventure theme park at Disneyland in California. Turning the corner and seeing the street extending away, I was briefly taken in. But this was quickly followed by the realization, "oh, it's just a picture." Wollheim himself argued that encountering *trompe l'oeil* paintings was not a case of "seeing in" because it lacked the twofold structure characteristic of picture seeing, and because of this, for him *trompe l'oeil* was not a pictorial phenomenon (Wollheim, 1987: 62). Perhaps VR involves a similar situation in which viewers are completely taken in and do not even notice the VR surface?

In principle, this would be possible, and such would be the case in the perfect virtual realities noted at the beginning of this book, but for the most part ignored here because of their tendency to distort the issues. However, in practice, given the prominent artifactuality of VR technology—the headset, the pixelized screen—I doubt that VR media could consistently give rise to complete illusionism, and I suspect the depictions in VR are probably much less susceptible to being mistaken for real objects and scenes than *trompe l'oeil* paintings are. But even if the VR surface were sometimes misperceived in this way, this would not disqualify VR as a case of picturing. The point of Feagin's example was not to show that the occasion was one in which her friend did not see the picture, but rather that sometimes when we see pictures and perceive their content, we misunderstand what we see. That such cases can be accepted as cases of twofold pictorial seeing is again to reassert my earlier claim that the relevant sense of "seeing" in pictorial seeing need not be awareness of, or attention to, the picture surface. Twofold theory, perhaps, needs to be understood in more minimal terms than Wollheim intended.

A point of method can be made here. Whether or not the surface is perceived as a pictorial surface might be thought to be beside the point of whether that surface is a picture: what is of relevance is that the surface is marked so as to provide opportunities for perceptual seeing of depicted objects, and that this surface actually does factor into the acts of object and scene perception that follow. Precisely what the nature of this seeing is in various cases, is a secondary issue which will contribute to our understanding of *how* these pictures function, but not of whether they are pictures or not. The discovery of a form of picturing—whether *trompe l'oeil* or VR—that departs from our current understanding of pictorial seeing does not undermine the pictorial status of that case, but rather is a problem for a theory of picturing that does not allow for it. We may simply need to adjust our theories in light of the new case.

VR illusionism, in one sense at least, is quite consistent with what we know about how the visual system attends to VR media. In the technology of VR there is a well-known problem called "accommodation/vergence decoupling" (Hoffman, et al., 2008). This phenomenon involves two features of the visual system key to object perception and noted earlier in the discussion of stereoscopic viewing. First, "For a stimulus to be sharply focused on the retina, the eye must be accommodated to a distance close to the focal distance of the object," but additionally, "For a stimulus to be seen as single (i.e., fused) rather than double, the eyes must be converged to a distance close to the object distance" (Hoffman, et al., 2008: 2–3). In ordinary experience, accommodation and convergence typically act in concert, but they "decouple" in the case of VR due to the first focusing on the screen, and the latter fixating on objects within the apparent VR space.[1] What this seems to show is that our visual system accommodates towards the shapes on the surface to bring them into focus, but that it also naturally converges on the apparent items in the picture space to pick them out for attention and locate them in egocentric space. To do so, "it is important for the graphic image to create a faithful impression of the 3D structure of the portrayed object or the scenes the displays depict" (Hoffman, et al., 2008: 01) suggesting that VR does rely on a kind of illusionism, at least regarding visual convergence. But the important conclusion here is that even if VR does involve a visuospatial illusion this does not establish that users do not see the screen and that it is not a case of pictorial seeing, rather, it explains *how* the user sees a visual scene through the screen surface: VR illusionism and the twofold seeing thesis, or at least the notion of *seeing in*, are thus consistent.

Finally, there is the ontological sounding claim that users of VR do not see screens but rather encounter VR worlds and objects directly. In recent work, David Chalmers makes the surprising claim the "in typical VR, one need have no sense of seeing a screen, and it can perhaps be argued that one does not see a screen at all" (2017: 319). Instead, what users perceive are "digital objects, constituted by computational processes on a computer" and that "virtual reality is a sort of genuine reality, virtual objects are real objects, and what goes on in virtual reality is truly real" (309). We could follow Chalmers and refer to this view as "virtual realism," but because I will retain this term for use in another context, it might be better to refer to it as virtual "ontological realism." We might also consider Grabarczyk and Pokropski's account to be a case of such ontological realism because their references to "beliefs" and "alternative places" (2016: 35) give the account a strongly metaphysical flavor that is out of step with twofold accounts.

I am going to leave aside until the last chapter these strongly metaphysical claims, however what can be said immediately here, is that if virtual reality media can be reconciled with the theory of pictorial "seeing in," this would do much to undercut these views; and given what we will find to be the metaphysical excesses of some accounts of VR, the theory presented here should be preferred over these accounts on the grounds of ontological parsimony. To explain how we perceive an apple in a still life painting, we need not consider the pictured apple as being real or as requiring our belief in its existence, and so if VR can be explained as a kind of picturing, albeit of an egocentric and interactive kind, we might avoid similar ontological claims concerning its apparent objects.

There is one final question regarding the pictorial surfaces of VR that I have often encountered or been asked: even if the current technology supporting VR involves pictorial surfaces, might there be ways of generating the experience of

VR worlds without such surfaces at all, and what might we think of those cases? Would that development render my approach obsolete? There has been a lot of talk recently, of course, about "implantable brain-machine interfaces," Elon Musk's Neuralink, and so on. The idea might be that, in the not-so-distant future, users of VR will encounter virtual realities not via a stereoscopic headset and peripherals, but through the direct feed of information into their brain from a computer simulation. This is clearly science fiction at the present time, and, of course, is largely equivalent to Gilbert Harman's brain in a vat hypothesis (1973). And it is not clear to me that this would undermine the project here, or affect any of my substantive conclusions, because this simply seems a different topic to the one treated here. There may be such future methods of conveying spatial experience, but it is not clear to me that these would continue to comprise *media*, instead counting as something like technologically manipulated hallucinations. These would no doubt raise their own questions, and for sure the prospect of such hallucinations might strengthen some of the ontological suspicions regarding VR experiences, but future developments do not undermine the current study, in as much as it deals with the current state of virtual media, which is itself of intrinsic interest. These are currently developments in picturing modes that seem to upset our previous understanding about the function of pictures, and that will remain the case even if these science fiction speculations come about.

What might we learn from surface-less VR? It is too early to say, because the assumption that such things are possible itself disguises any number of additional assumptions about the nature of the conscious experience of the world, and this is something that is not only a difficult scientific question, but also a perplexing philosophical one. Indeed, assumptions about what VR media itself is currently achieving and can achieve, open the way for a worrying possibility for VR boosters. It might be that successful surface-less VR would count, not as a virtualization of native experience, but instead as a *virtualization of pictorial technology*. That is, that it would count simply as a virtualization (in the sense defined in Chapter Two here) of VR picturing. I have assumed in this book so far, of course, that VR technology is a remediation of native experience, but this itself may be a faulty assumption. It could instead be that VR media, like pictures, comprise an attempt to convey an impression of perceptual experience, but that this attempt fails in a fundamental way, because we misconceive or simply do not understand what it is to experience the world and thus what it is that we are trying to remediate. If linear perspective was a failed attempt to remediate spatial experience, as some of its critics contend, then the aspirations of VR to again take up this remediation may ultimately fail, leaving only the conjectures of artists and philosophers about perfect virtual worlds.

4.4 Egocentric Picturing

VR currently does involve a screen, and users must see this screen even as they see through it to the items or scene that is depicted. But how is it possible for this act of pictorial seeing to place the viewer, *pace* Nanay, into an egocentric relationship with the items viewed, and in a way that the user might subsequently interact with those items? We can work up to this understanding, and in so doing learn what egocentric seeing comprises, by exploring what might seem, at first blush, to be a counterexample to the idea that traditional pictures do not involve egocentric seeing. Earlier I discussed how the technique of linear perspective implies a *point of view* on the pictured space. It is a short step from this to taking this point of view to be a *viewer*; and then, perhaps a further step to invoking the picture viewer as *the viewer of the scene*. More than just providing an apparent perspective on the pictured scene, some art has attempted to *implicate* the viewer within that scene. In his work on pictorial seeing and painting as an artform, Wollheim notes that there

> are paintings in which the suitable spectator is offered a distinctive form of access through the presence in the represented space—though not in that part of it which is represented—of a figure, whom I call the 'Spectator of the Picture'.
>
> (1998: 225)

There is a visual trope in painting where the viewer is depicted as inhabiting or at least standing in relation to the picture space. As mentioned, there is a hint of the technique in Raphael's painting, but other works go much further in acknowledging the presence of a viewer of the scene beyond the picture space. Jan van Eyck's *Arnolfini Portrait* and Diego Velázquez' *Las Meninas* both exploit this implication to place the viewer in an apparent relationship to the action in their pictorial spaces. In van Eyck's *Arnolfini Portrait* (1434) two people—one of whom may be Jan van Eyck himself—can be seen reflected in a mirror on the back wall, in a way that spatially implies that they are the viewers of the scene shown in the painting, perhaps occupying the position of the actual viewer. The effect is not precise because van Eyck could not employ strict linear perspective in the painting, but the impression of a viewer just beyond the pictured space is given, and partly because the figures in the painting present themselves to this apparent viewer. Velázquez' *Las Meninas* (1656) is another example where the apparent viewer of the scene stands just outside of the picture space, located where the gazes of several of the depicted individuals converge. In this example, the viewer of the scene may be identified with King Philip IV of Spain or his wife. A further intriguing complication of this painting is that the artist Velázquez is depicted as looking to this couple, and presumably sketching or painting them on a large canvass before him that may itself—though this is disputed—be reflected in a mirror on the far wall. Nevertheless, both paintings incorporate the apparent orientation on the picture space of the viewer themselves.

This gives us the background to understand part of the artistic achievements of the *Arnolfini Portrait* and *Las Meninas*. Both paintings extrapolate from formal developments of perspective—in the case of *Las Meninas*, strictly linear perspective and in the *Arnolfini Portrait* a less rigorous sense of space—to realize an artistic technique. In both cases, the depicted space of the painting implies the presence of a viewer within the picture space, and moreover, implies that this viewer occupies the space of the actual viewer. Both works thus connect the viewer with the apparent scene viewed, relating the viewer with the space "in front" of them so that they can become a part of the implied content of the work. Moreover, both works employ this arrangement to develop subtle meanings. In the *Arnolfini Portrait*, the depiction of space and the inference that a viewer of that space exists beyond the picture plane, is exploited to considerable artistic effect. The meaning of the *Arnolfini Portrait* is much contested, but it has been famously (though just a little controversially) suggested by Panofsky that the viewer—whose presence is depicted in the mirror, but also implied by the pictorial space of the painting—has the role of a witness to a marriage which is depicted in the scene (Panofsky, 1934). The meaning of the painting, then, can only be grasped by understanding the spatial implications of the work.

However, if we consider these pictures more closely, and just how they implicate the viewer in the scene, we can see that they are not genuinely egocentric pictures. Like linear perspective itself, the spatial effect of the "implied viewer" or "spectator of the picture," is problematic, and gives only a rough approximation of egocentric experiential spaces. If the viewer of *Las Meninas* moves before the painting, unlike our experience of a real visual scene, the apparent relative position of the items within the picture space do not shift to respect the viewer's change in orientation. Moving toward the painting does not bring one closer to some figures within the picture relative to others; moving to the side does not reveal what was previously occluded by figures in the mid or foreground of the picture. While it gives an impression of the viewer's position, *Las Meninas* does not imply that the scene is viewed from any specific location congruent with the picture space, but rather, that the scene is merely *before* the viewer because the painting itself is before the viewer. Genuine "egocentric" viewing of the kind seen in VR requires even more than the technique of linear perspective and the implied viewer.

This egocentric relationship between the picture viewer and picture space in VR is best understood as the kind of counterfactual tracking relationship between the implied perspective and the *actual* perspective of the picture viewer that was introduced in the previous chapter and that I have discussed elsewhere (Tavinor, 2019a: 148). In a truly egocentric picture, if the viewer moved their head to the left, then the contents of the scene would appear as though viewed from that new position. But this is not what happens in a picture such as *Las Meninas*, because if the viewer of that painting moves to their left, the implied position of the viewer within the picture space remains static. The viewer of *Las Meninas* can in fact move quite freely before the picture and still receive the visual impression that the space exists before them, their perspective on the space unchanged. It is partly to explain why the viewer's movement before the pictures does not distort the scene that Nanay distinguishes the roles of the ventral and dorsal streams in pictorial seeing (2008: 977). Note also that this fixed and non-counterfactual dependence of the picture view and apparent view in painting is also the reason why the eyes of the figures in such pictures tend to follow you, and why, *pace* Klevjer, this is not evidence for the egocentric nature of a picture. If *Las Meninas* really was egocentric, the fixed eyes of the Infanta Margaret Theresa would *not* follow you as you moved before the painting, but rather would be fixed on the space you previously occupied. It seems then, that the viewer is implicated in the work, but only in a very loose way. Hence the problem is that in *Las Meninas* the implied position of the viewer of the scene does not track—is not counterfactually dependent on—the actual position of the viewer of the painting.

The central trick of VR picturing is to allow the apparent position of the scene viewer and the actual position of the picture viewer to bear this counterfactual relationship. This trick is achieved by a combination of the interactive structure of a 3D picturing environment, the tracking of the user's movements and projection of these "into" the picturing space, and the stereopsis of a VR headset. To summarize: in VR, the surface—typically patterns on a pixel array viewed through a pair of lenses—are generated via algorithmic means as a computer model of an environment rendered on a screen. As in conventional picturing, elements of the 2D surface—borders, light gradients, occlusion shapes, and so on—act as triggers or markers for the perception of 3D config-urations through the picture plane (Nanay, 2008; Hyman, 2006). But in VR, in addition to these features of conventional picturing and picture viewing, the production of the sense of spatiality in the picture also involves the tracking of the user's body movements into the 3D picture space so that when they move, the orientation of the virtual camera changes to represent that movement. This tracking is the core of the counterfactual dependence of VR picture viewing as it allows the picture content to change with respect to the viewer's actual position and orientation. Finally, VR exploits the native stereopsis of the human visual system to place this tracked perspective in the depictive space. The binocularity of VR stereoscopic headsets gives a naturalistic impression of depth to the 3D models, and as a result the visual system interprets the resulting "objects" as having a very precise apparent spatial position relative to the apparent position of the user. Moreover, in VR stereopsis, the eyes converge *into* the simulated space beyond the picture plane, showing in a particularly robust way, that as far as binocular vision is concerned, VR picture viewing really is egocentric.

Earlier I speculated on the potential for a VR depictive surface which I called "Alberti's virtual window" that could render a 2D pictorial plane cor-responding to the tracked viewer's position, so that they would seem to look into a virtual space. Let us revisit this case of virtual *trompe l'oeil* and see if that also counts as a case of egocentric picturing. This case would differ to stereoscopic viewing in that the viewer, though related to the pictorial space, would be unable to move into it, restricted to viewing from the threshold (perhaps then, the technique might also be referred to as "Alberti's virtual doorstep"). The viewer also, like other cases of *trompe l'oeil*, would be unable to interact with items in the depictive space. Nevertheless, the technique also allows the space "beyond" the screen to alter to display the geometrical configuration appropriate to egocentric viewing, that is, it allows for a counterfactual relationship between the movement of the viewer and the apparent positions of items in the depicted space.

Employing both stereoscopic viewing and Alberti's virtual window allows for a mutual dependence of the viewer's actual spatial orientation and their apparent orientation in respect to the scene depicted. But we should take care to note that how the actual orientation of the viewer is tracked differs in each case. In Alberti's window the viewer's orientation is tracked relative to the screen surface itself, so that their movement in front of the screen alters the depictive space projecting away from them through the picture plane. In the case of a stereoscopic headset, the viewer's orientation is not tracked relative to the screen, which after all is always fixed directly in front of the viewer's eyes; rather, the user's bodily movements are tracked within the space they occupy. It is this orientation that is depicted on the screen and projected into the depictive space of the 3D environmental model to give the visual impression that the viewer themself occupies a defined but moving location in that perceptual space.

But we are now presented with a worry about the idea of egocentric picturing: what this looks like is not a theory of egocentric picturing, but of *apparent* egocentric picturing. Above, when I discussed the role of stereopsis in VR picture perception, I noted the "eyes converge *into* the simulated space beyond the picture plane," where, in fact, there is no corresponding object or space to be seen. In the case of *The London Heist*, of course, it is merely fictional that one is located in a pub, and so one might say that it is merely fictional—or simulated—that one is located relative to the depictive content of the picture. The viewer in this case of VR seeing is not really in a spatial relationship to a perceived object at all. This, the critic might say, shows that there is no real egocentric viewing occurring in such cases, but only apparent, fictional, or simulated egocentric seeing.

However, this would be a mistake because it ignores the very fact that we are discussing picture perception, where the lack of a real and present object corresponding to the object that is pictured is probably the typical case. Any theory of what would count as egocentric *picture* viewing—rather than egocentric native viewing—should presume that pictured objects are often, if not typically, not present to the viewer, or even real objects at all.[2] When you view a still life painting on the wall of a gallery, of course, there really is not an apple in front of you, rather, it is likely fictional that this is the case. The theory of egocentric picturing presented here aims to discriminate between the picture viewing of such still life depictions of apples which are not egocentric, and cases such as VR which apparently are egocentric, even though in both cases there may be no actual apple that is being perceived and hence no real spatial relationship, egocentric or otherwise, between viewer and object.

How does Wollheim's "spectator of the picture" fit into this account of egocentric picture viewing? Is the apparent spatial perspective in egocentric picturing—the virtual location and orientation in graphical space that generates the 2D picture plane—necessarily associated with an implied viewer? I began this section with discussion of the *Arnolfini Portrait* and *Las Meninas*, pictures in which the apparent viewpoint is associated with a specific viewer, whose presence might even be pictured (as reflected in a mirror, for example). Does VR necessarily imply the presence of this viewer? In many, perhaps most cases of VR picturing, the implied viewer does exist, and moreover, it is implied that the user is to be identified (perhaps fictionally) with the viewer. However, the existence of the virtual perspective does not always mean that this perspective need be associated with a depicted agent. In many VR experiences, such as portions of *Apollo 11 VR* (2016) the user's virtual perspective is a disembodied viewpoint on the action, more like a simple virtual camera than an implied viewer. In this VR experience, while the user is able to take the controls of the lunar lander, in many of the VR sequences the user is merely given an apparent spatial perspective on the events of the Apollo 11 mission to the Moon in a way that no actual spectator could have viewed them. While it seems especially natural to embody the VR user as a spectator (and actor) in the position defined by the apparent perspective because this replicates the situation in our natural experience of the world, this is not inevitable. This issue takes us quite close to the question of the role of fiction and the imagination in VR, and the question of whether users of VR imagine themselves to be viewers of the scenes pictured, something I will cover in the next chapter.

4.5 Virtual Pictures and Interactivity

The final issue that might lead us to conclude that virtual visual media are not in fact a form of picturing can now be addressed. Virtual media, it might be thought, are quite unlike pictures because they are *interactive*. As noted, in his account of pictorial seeing, Nanay claims that "we don't and can't perform actions on depicted objects" (2018: 189). Klevjer similarly claims that "Depictive images [...] do *not* recognize agency as relevant, because they are not present [to the viewer]" (2017: 21, emphasis in original). And yet, it is credible that VR users can "perform actions on depicted objects" and that the medium does "recognize their agency," in so far as players of VR games such as *Half-Life: Alyx* may, as I claimed earlier, "physically gesture, reach out, and interact with the objects and characters" of their worlds.

There are, of course, limits on VR interactivity, and so we should note that not all VR applications allow for much in the way of interaction, and all VR applications are essentially limited in their interactive potential. VR games like *Half-Life: Alyx*, despite their often-surprising interactivity, are still relatively restricted interactive spaces. Of course, this need be no great difficulty for the representation of gameworlds in VR, because games always constrain possible play actions, limiting them to what the philosopher Bernard Suits refers to as

"constitutive rules" (Suits, 2014). It is also worth raising here the loose terminological variation between VR games and VR *experiences*. VR games involve
the necessary range of interaction to allow the functioning of the game; but in
VR experiences, the user's agency is particularly constrained, often to merely
looking around at what is basically an egocentric VR film. For example, in the
various VR theme park virtual movies and dark rides, while the orientation of
the user—in this case, rider—is within the depicted space, they cannot effect
any changes of the depictive content, as it is, often literally, "on rails." The 360-
degree videos briefly mentioned earlier are also typically of this non-interactive
form. This distinction between interactive and non-interactive VR experiences
is also evident in VR art, with some such art allowing the audience to change as
well as view the virtually depicted or sculpted surface, and other virtual art
involving a passive viewing experience. But even looking around and directing
one's attention to different parts of the VR graphical space is a kind of agency,
and so VR experiences and non-interactive VR art are nevertheless interactive
in this limited sense. But how is VR interactivity possible at all? What is it for
the user of VR to interact with the objects depicted by the medium?

There is a simple-looking response here that should be avoided because
though it contains some truth, it is not simple at all. This is the idea that
users of VR simply interact with virtual apples and other such *virtual
objects*. Phenomenologically this response might seem credible, especially if
we take as our exemplar of VR a hypothetical perfect virtual reality. Under
such a case, it would really seem like you were simply picking up and perhaps even biting into an apple. And some theorists of VR, as I have already
suggested, grasp this simple response with both hands when they adopt the
position of virtual ontological realism. I do not find this position credible
for reasons I will explain in the final chapter. But even if we restrict ourselves to actual VR media, in a couple of senses at least, there is some truth
to the idea that we interact with "VR objects." First, on the occasions in VR
when a person virtually gestures toward or grasps an apple, that action
surely involves interacting with *something*, and we might out of convenience
call this the virtual object. I will try to explain here what that something is,
and the obvious candidates are the various physical aspects of the computer
technology underlying VR described in the previous chapter. Second, in
some limited cases we will find that VR allows for real interaction with the
objects it depicts, and so in special cases it might be that VR users genuinely
interact with depicted apples. To navigate these issues will take some care,
and again, will involve avoiding the distraction of perfect virtual reality, and
carefully inspecting the actual media and their use.

So, how can pictures sustain user interaction in a way that explains the
actual interactive phenomenology of VR games, art, and experiences? There
are two parts to this story. The first part concerns the interactivity of the
medium of VR, that is, how the computational medium standing behind VR
registers and incorporates the interaction of its users. The second aspect of
the theory concerns how interactivity of the pictorial medium of VR gives

the impression that the user interacts with the depicted items. That is, how in VR media the user is pictured as picking up and manipulating objects such as guns, mobile phones or apples.

The first thing we need to understand is that VR, like some other media, are *interactive*. What is it then, for a medium to be an interactive one? The words "interactive" and "interactivity" have been used with such a profusion in recent times, that we might suspect they, like "virtual," are buzzwords that refer to some aspect of our current technological predicament, but in an imprecise and merely suggestive way. In fact, interactivity has shown signs of going out of fashion very recently, with virtuality increasingly prominent in this role.[3] Nevertheless, there is a philosophically respectable and useful sense of the term "interactive." Medium interactivity has been the subject of some interest in aesthetics, driven by the increased interest in computer art and videogames (Lopes, 2010; Gaut, 2010; Tavinor, 2009). While philosophers are not in total agreement about the nature of interactivity[4] it is credible that interactive works are those that "prescribe that the actions of [their] users help generate [their] displays" (Lopes, 2010: 36) or cases in which the work "authorizes that its *audience's* actions partly determine its instances and their features" (Gaut, 2010: 143, italics in original). In this sense, the videogame *Super Mario Bros.* (1985) is interactive because exactly what is displayed on the screen, though largely shaped and constrained by the game's design, is determined by how the game is played. For example, precisely when Mario jumps on a goomba is controlled by the player. But the notorious film version, *Super Mario Bros.* (1993), is not interactive because as a viewer you sadly have no influence on the content of the movie, which has been largely fixed at the time of production of the film.

Virtual media are typically interactive in this sense, in that the user, who is at the same time the principal audience of the interaction, can almost always alter what is displayed in a given work. VR "experiences" are minimally interactive in this sense, in allowing the user only to look around them at a largely scripted experience or story, and thus to spectate only; but in most VR applications, including all games that utilize the depictive medium, the user can perform actions that alter the content displayed by the depictive surface, and hence, what is seen through this surface. In such cases the spectator is also an actor, and in its visual modes, at least, VR is best considered as a kind of interactive picturing.

Photographs and paintings are not typically interactive in this sense, in that they do not have an interactive potential such that the audience can interact with, and change, their depictive content. However, some traditional pictures do allow for interaction, and these exceptions tell us something about the technical means of the interactivity of VR. For example, the pictures in a pop-up book are, in this sense, interactive, but are so in a purely mechanical way in that they are physically articulated in a way that supports user interaction. This mechanical articulation allows the user to alter what is depicted in the picture by changing the surface, such as where one might open a cardboard door to reveal the occupant of a house or pull a tab so that a dog runs across a field. In such cases, the user directly manipulates the surface to change what is marked on, and can be seen through, the picture. VR media are similarly articulated, but this time in an *algorithmic* rather than mechanical way. In VR, the technical means by which the user alters the display is an algorithm that generates a pictorial surface resulting from their input into a controller, or how they move the tracked surface of their own body. In principle, though not in the material means and scope of the interaction available, the two cases are equivalent.

We might also conclude that there are two folds of interactivity inherent in VR picturing, relating to the user's ability to change the *framing* and *content* of the picture. When Raphael was painting his fresco on the walls of the Apostolic apartments, the content of the work and the perspective on that content were simultaneously fixed because these are both bound up in the production of the image surface. Similarly, in photography, the scene and perspective are bound together by a mechanical and chemical process at the time the picture is taken. But in computer graphics, the independence of the 3D model and the virtual camera, and the interactivity of both, mean that a user can independently modify the depicted content or the perspective on it as they are rendered. In VR experiences and many VR artworks, the interactive framing of VR allows for explorative acts of picture perception in VR. When the modelled objects and environments themselves can be manipulated, the user has an influence on the objects or events depicted. This allows for direct creative intervention by the viewers that is exceedingly rare in previous pictorial traditions, as we will find.

But the interactivity of VR pictorial media, so characterized, quickly brings to our attention the second sense of interactivity listed above: that is, the apparent interaction that users have with the *objects* depicted within VR. Let us put to one side until the next chapter the considerable complication of what these pictorial objects might depict beyond the picture and concentrate on the objects *apparent* in the pictures themselves. It should be uncontroversial that interacting with VR media often gives the impression that the user or player interacts with these objects. The language that users of VR employ often involves the first-person reference to their interaction with such objects: a player of *Tumble VR* may report that they "dropped the cylindrical piece, bringing the puzzle tumbling down." This apparent interaction is something altogether different from manipulating the pictures in a pop-up book, because in the latter case, the user or their actions are not a part of what is depicted: one might open the door in a pop-up book to reveal what is inside, but the pop-up book does not itself depict the user performing this action.

But the situation is quite different in interactive media such as video-games. In videogames, the interaction with the pictorial medium is semi-technically described as involving the interaction with a *prop* triggering an *event*. A prop might be almost any kind of videogame *asset* such as a 3D model, and when interacted with, usually by the press of a button, the resulting event is usually itself animated. So, to take a very simple example, opening a door is a common interactive event in videogames, and involves a prop with an interactive affordance that can be used to trigger an event. When the event is triggered, the screen might depict the hand of the player-character reaching toward the door and pushing it, and the animation of the door opening to reveal a space beyond. Thus here, the result of the interaction itself is pictorially configured.

In a typical videogame, all this occurs via the medium of a 2D screen, and so the "hand" that reaches toward the door, given the spatial limitations of 2D picturing, gives only a vague impression of reaching forward from the position of the player. But because of its egocentric nature VR adds a more precise level of spatiality to this apparent interaction. An example from *The London Heist* is worth discussing in a little more depth here. In one scene, set in the dingy English pub, a mobile phone sits on a table before the player, and it is natural to reach out and pick it up as a text notification comes through. In a later sequence, a character hands you the phone to take a call. Again, I found myself reaching to take the phone, placing it to my ear, and then hearing a voice from the other end localized in the ear to which it was virtually being held. The naturalness of this interaction was surprising to me as a recent initiate to VR.

This interaction gives an altogether stronger impression that one is interacting with the objects depicted, because as argued earlier in this chapter, the egocentric picturing in VR allows the user to be spatially related in picture space to the pictorial objects. And not only can users perceive the apparent spatial relationship they bear to pictured items, but based on this spatial relationship, they can reach out to and interact with these items, producing additional content that is pictorially rendered contingent on their interaction, in this case, a graphical rendering of the answering of the phone. In the case of *The London Heist*, triggering the interaction with the phone still involves pressing a button of the controller that is tracked in the player's hand, but in other cases the interaction might be initiated by grasping the apparent object while wearing a tracked haptic glove. This mode of interaction is what was earlier referred to as "gestural" control. Hence, VR involves not just tracking a user's perceptual orientation on pictorial space, but also potentially, the tracking of their bodily position, movement, and agency into this space so that they can generate interactive events there. The tracking of vision and action must be spatially congruent if the gesture is to be successful in locating the relevant position in pictorial space, and if we are to consider the vision and action to be in one virtual space.

In some VR applications the depictions arising from the user's interaction may involve the picturing of hands, or a tool or weapon, itself being represented within the 3D graphical scene. In the *London Heist* mobile phone example, the player's interaction is depicted in the form of a leather glove manipulating objects in the gameworld. In other games and apps, however, the interaction is not pictorially realized, giving the impression of a disembodied interaction. The difference between the two cases perhaps regards the detail of the representation of the "spectator of the picture" that Wollheim contends to be a feature of many traditional paintings. The former case is richer in its depiction of the spectator's body as having a presence and location in the world—such as in the non-egocentric case of the *Arnolfini Portrait* where the viewer may be depicted in the mirror—whereas in the disembodied case this spatial presence of the body is merely implied. Nevertheless, this seems to expand on the potential of the depicted spectator seen in the *Arnolfini Portrait*: not only can the spectator be seen in a mirror, but the user's apparent hands, body and tools can be inspected from the pictorial first-person. This is even if, as in the case of *The London Heist*, the player is depicted without arms, resulting in a spooky floating hand moving through the space grasping items before them.

So, if we take this interaction—the user interacting with an algorithmically generated picture via a control surface or tracked movements, and thereby generating a representation of their own apparent interaction with the virtual picture objects—as being one where the user can affect the objects in the scene, then *pace* Nanay, VR pictures do allow users to "perform actions on depicted objects." Of course, such pictorial interactions might be quite different to interacting with actual objects in the real and present world. But given that Nanay already accepts that *seeing* depicted objects is quite different to seeing actual objects "face to face" (2018), this should be no bar to admitting this to be a genuine case of interacting with depicted objects. Again, any theory of what would count as interactive *picture* viewing should presume that pictured objects are often, if not typically, not present to the viewer, or even real objects at all.

And hence, I see no reason to deny that stereoscopic VR media are a picturing technology on either of the grounds put forward by Nanay. The picturing surface exists in VR, and it is seen by the viewer (even if it is not always attended to), through this surface users perceive a depicted space, and finally, the medium affords users a means of interacting with depicted objects. We should subsequently conclude that VR pictures do locate the depicted objects in the viewer's egocentric space, and if Nanay's claims about its role are otherwise true, that the dorsal stream *is* involved in the perception of virtual scenes (and as we saw earlier, the *convergence* of the visual system certainly is). Returning to *trompe l'oeil* paintings, we might conclude that Nanay would be comfortable with this partial counterexample to his theory. In his conclusion to a paper that focuses on *trompe l'oeil* paintings and pictorial seeing, Nanay accepts that such paintings are a case of picturing, but uncharacteristically are one in which "our dorsal stream also represents the depicted objects" (2015: 193). He concludes that the consideration of *trompe l'oeil* adds to our understanding of important variations in picture perception.

This is also a benefit of investigating VR as a form of picturing because the present investigation allows us to see how egocentric picturing can be utilized in interactive ways. In *trompe l'oeil* the illusion requires the viewer to view the picture from a defined, stationary, and inactive perspective, but in VR, the technology of the medium—including the computer models used to generate a 3D world, the stereoscopy of the headset, and the motion tracking that allows the display to accommodate the user's changing perspective on the world, and their gestures within it—allows the spatial scene to be encompassing, dynamic and interactive. Where reaching out to interact with an apple in a *trompe l'oeil* painting would lead to a kind of "mistake," VR incorporates the interactivity of its computational medium so that gesturing toward or reaching out to a virtually pictured apple need not count as a mistake. As such, egocentric VR picturing may remediate our agency. But for what point or purpose? Pictures are often a means of seeing things *beyond* their surfaces, whether that be cities on the other side of the world or people of the distant past. What, *ultimately*, is being interacted with when one interacts with a VR object? This, as we will find in the next chapter, depends on the use to which the VR medium is put on a particular occasion.

Notes

1   This decoupling of converged binocular vision and the accommodation of the focal point of the lens on the screen is a significant problem for VR design, as it can cause fatigue and headaches in users, perhaps contributing to so-called VR sickness (Hoffman, et al., 2008).
2   However, in the next chapter I am going to discuss a non-typical case where the picture apple may be real and present to the viewer.
3   Google's Ngram Viewer shows the word frequency of "interactive" to have peaked just before the turn of the century; the peak usage of "virtual" is much more recent (and one suspects its usage will soar as Ngram begins to take account of the impact of 2020 on the word's use).
4   See for example the alternative accounts of interactivity provided in Smuts, 2009 and Wildman and Woodward, 2018.

5 Seeing and Doing with VR Media

5.1 The Contexts and Uses of Virtual Media

In the previous chapter, in explaining the egocentric and interactive nature of VR media, I concentrated, quite understandably, on their prevalent use in the fictions of typical videogames. However, a recent set of experiments gives a surprisingly different sense of the potential of virtual reality media, and the capacity of egocentric picturing to support perception and interaction. In Mathew Pan and Günter Niemeyer's work at Disney Research Studios (2017) a user wearing a stereoscopic headset is able catch a real ball that is tossed to them, based entirely on the motion-tracked ball's depicted movement in a VR environment.[1] Employing a motion tracking camera, a ball's position is tracked and displayed on the stereoscopic headset as an animated ball within a rudimentary virtual environment comprising a textured floor, basic lighting, and paddle-like depictions of the user's hands (Pan and Niemeyer, 2017: 269). Using this visual information, the user can orientate her hands to achieve the task of catching the ball in virtual and in real space.

This is a surprising use of virtual technology because the remediation of spatial experience here is used to interact not with the zombies, aircraft, or race cars familiar from the worlds of videogames, but with the furniture of the actual, real, *present* world. Specifically, VR in this case is used to *see and catch a real ball*. This might be a surprising instance of virtual reality, because we so frequently take that term—falsely, as I argued—to be contrasted with reality itself: virtual worlds, and our interaction with them are not real, but rather, merely imaginary, conceptual, or such like. But as we will see in this chapter, the use of egocentric and interactive VR picturing technology to support real interaction in the real world is perfectly fitting with the theory of virtuality as a non-customary remediation of the functional efficiency of everyday seeing and agency. The ball-catching example and others like it provide further evidence for the correctness of the analysis of VR offered here.

To see this, let us return to the question posed, and left half-answered, at the end of the previous chapter: what is it that is interacted with in the case of VR agency? I argued that the interactivity of VR is based first on the interactivity of the medium, and second, on the interactivity of the apparent 3D objects depicted

DOI: 10.4324/9781003107644-5

in virtual space. However, when we consider the contexts in which VR is employed, the question of what is interacted with expands and changes focus. It turns out that there are multiple kinds of objects with which the user of VR can interact or at least appear to interact. The medium of VR allows users to interact with and change, via algorithmic means, elements of a pictorial surface; but through this, it can also allow a user to see and catch a real ball that is travelling toward them; fire a gun at the encroaching zombie hoard; see the imperial palaces on the Palatine Hill in Ancient Rome; remotely view the surface of another planet; or learn about medical techniques they will later employ in their working life as a health professional. Of course, traditional pictures themselves have a variety of uses, so we should not be surprised that this new form of egocentric picturing itself has multiple practical implementations. Pictures can be used to provide the content used to ground flights of imagination; to represent ideas in iconic form; to document events in the real world; to instruct and inform; and to see one's great grandparents in photos on the wall. The use of the egocentric and interactive pictures of VR largely follows the use of traditional pictures, though with variations associated with their egocentric and interactive form.

To appreciate the variation in the pictured objects in VR, we need to revisit the ideas of "twofold seeing" and "seeing in." As we found earlier, the picture surface is the first direct object of perception, and we perceive it, and sometimes attend to it, as marks on a 2D plane. The second fold comprises the spatial configurations or three-dimensional objects that might be recognized in the pictorial surface, because the surface comprises a color field that encodes features like borders, occlusion shapes, and textural gradients, that our visual system perceives as spatial configurations. But in addition to this act of picture perception, we may understand the picture viewer as also seeing what the picture is a *picture of*, be it Aristotle, New York City, or a zombie hoard rushing toward you. We might call this the "intentional object" of the picture as a way of distinguishing it from the pictorial object of the 3D configuration inherent in the pictorial surface.[2] Successfully perceiving what a picture is a *picture of* is not something that can be directly perceived from the picture itself but depends on the contextual understanding of the viewer of the kind of picture they are viewing. This is because, first, what a depiction is a *picture of* may not be entirely apparent in the picture itself. The 3D pictorial object in Picasso's 1910 cubist portrait of Ambroise Vollard does not especially look like Ambroise Vollard, even though it depicts him. Secondly, two identical pictorial configurations might ultimately depict different things. Two identical paintings, to use a well-worn example in aesthetics, might depict alternately, an historical king, or a person in fancy dress. The difference between the two paintings must be something about their intentional context and historical provenance: namely, that one was painted to depict the king, and the other to depict our friend Trevor who is fond of fancy dress. Much of the utility in the picture derives from this broader notion of pictorial representation and recognition: it is because a picture may be of a real person, that we may hang pictures of great grandparents in photos on the wall.

What this analysis of the extra representational dimension in pictorial seeing allows us to see about VR is that the medium itself may have some constancy over its various uses, that is, it may involve configurational and recognitional seeing in all cases, but then the "intentional object" may differ considerably given the actual use of the medium. Similar to the example directly above, it is conceivable that the configured objects of a VR flight simulator—an aircraft and its environmental situation—may depict *fictional* or *actual* flight depending on the intentional context. This extra level of articulation in the theory of pictorial seeing hence allows us to distinguish a further range of uses of VR. These uses may in some cases involve the additional recognition of an actual person or place; in others, it will involve the recognition of a fictional character that is to become the basis of imaginative participation; in others it could involve the recognition of a ball that is travelling very quickly in the user's direction; and further still, a VR picture might resist being seen as having an intentional object beyond the pictorial configuration, being as it encourages us to involve ourselves with the surface and its configuration only, perhaps because the image primarily calls for aesthetic interest in the surface. These varied uses should not come as a surprise, in that VR comprises the remediation of experience and agency, and this remediation is useful in more than one context. The following is far from an exhaustive survey of the uses of VR—given the size of the field, that would be impossible—rather it is a selection of uses of VR that identifies several crucial contexts for the production and viewing of egocentric and interactive VR media.

5.2 Virtual Fictionalism

An initial use of VR picturing to examine, and one that is perhaps the most widespread case, is in the service of fiction. In videogames, VR is typically used to provide content for the purposes of make-believe or the imagination. Any number of examples can be used to illustrate the fictional content of VR games. In *Half Life: Alyx*, the player, in the guise of Alyx Vance, a young woman, battles the Combine, a race of interdimensional aliens who have invaded the earth. In *Resident Evil VII*, the player, as the role of Ethan Winters, searches the dark mysterious house for his missing wife. In *Minecraft VR*, various fictional goals are available for the player: building a house to survive the first perilous night, mining for diamonds, protecting your villagers from the invading pillagers, or killing the Ender Dragon. All these activities are fictions because they engage the user's imagination and because the configurations of the pictorial surface depict objects and events with an imagined existence. Thus, without doubt, many of the uses of VR involve the depiction of fictional content.

Now, a contrary question here might be, given the details of the previous description, Is not all VR fictional? Aren't all the instances of VR described in this book, and the way they are described as being users seeing objects in a pictorial surface, cases of *fictionally seeing objects*? I can imagine, how on

Kendall Walton's view of pictorial seeing, they might seem to be, in virtue of VR counting as a kind of representational device that involves its users adopting the attitude of make-believe. Walton argues that twofold seeing, in its second fold involves the pictorial viewer imagining that they see a given object or scene (Walton, 1990: 293–297; 2002). In this theory, the first fold is seeing the picture surface, and the second fold is the imagining of this perceptual act, that it is seeing the object depicted. Pictorial seeing is thus an act of the imagination, and on this view, it might seem that seeing anything in VR—the puzzle pieces of *Tumble VR*, Alyx's perspective in her fight with the Combine, a view of Ancient Rome—is a fictional act of seeing. Under this view, even recognizing the configured content of a VR depiction would count as an act of imagination, implying perhaps that it is fictional that any seeing of scenes, objects and figures occurs in the case of VR.

The proper response here, I believe, is settling what Wollheim refers to as the appropriate "division of labor" between perception and the imagination (Wollheim, 1998: 225). The imagination surely is involved in many uses of VR, as I will set out here, particularly when the viewer is the "spectator of the picture" or "implied viewer," and most obviously, where VR involves and represents its participants in adventures in fictional roles such as Alyx Vance and Ethan Winters. But these occur only after the perceptual process of "seeing-in" has done its job and allowed the user of VR to recognize the content configured in the pictorial surface. I suspect, like Nanay, that the work of picture perception is already done before Walton's imagined viewer enters the scene (Nanay, 2004: 286). I do not deny that VR sometimes involves the imagination, and that it results in fictions about the user, what they see, and the actions they take. What I deny is that *seeing in* is a result of this imaginative involvement. Hence, I do not see the basic account of VR presented here as one of *virtual fictionalism*, even though some cases of VR involvement are properly referred to as virtual fictionalism. VR is often used as a prop for the imagination, but on other occasions, the picturing technology of VR is employed for documentary, factual or even prosthetic purposes. Let us see what kind of fictive cases there are, and how we can distinguish them from the other uses of VR.

Half-Life: Alyx and *Resident Evil* are both easily seen as fictions because they are so close in form, in many respects at least, to other undisputed fictions such as cinema and television fiction. They involve fictional characters and environments, a plot, and happily fit into the familiar fictive genres of science fiction and horror, each drawing on the tropes and techniques of those genres. In both we find terrifying monsters, jump scares, and the story of a protagonist battling against overwhelming odds. But other cases of VR, potentially lacking features we associate with cinematic or television drama, are also fictions because of what they depict. *Microsoft Flight Simulator*, while obviously a flight simulator, presents the fiction that one is flying an aircraft. *Apollo 11 VR*, at least in the section where you fly the lunar lander, presents a fiction of traveling to the moon (even as other sections of the experience may merely document aspects of the actual Apollo

11 mission). Even medical simulations, of which there are now many, sometimes present a fictional scenario. While they are used for training purposes, what one "does" in the simulation—checking a patient's vital stats, putting an intravenous line into a virtual patient—is fictional. Another interesting and increasingly widespread use of VR fiction is in the testing of scientific hypotheses, a practice which has led to some weird and wonderful examples of the medium. One 2020 study utilizing VR measured the appetites of people at a virtual buffet (Cheah, et al. 2020). One could design an actual experiment to document food selection amongst emerging adults at a buffet, but a fictional buffet is more convenient and cost-effective (though much less delicious). Incidentally, this study found that "participants' food selections in the VR and RW [real world] food buffets were significantly and positively correlated" (2020: 1).

While differing in what they depict—flying an aircraft, treating a patient, eating too much food at a buffet—and for what purpose, all these uses depict events, actions, and scenarios which are merely imagined to be the case. In such uses, the apparent intentional objects of the depiction comprise fictional objects: in *Half Life: Alyx*, for example, these are imagined entities such as Combine soldiers, gravity guns, and head crabs. I am not committed to any specific theory of fiction in my presentation of this account, and I think my position on VR will be consistent with many such accounts. So, for example, following Walton, we might claim that in most current uses, VR media act as fictive props in games of make-believe (Walton, 1990). The egocentric pictures and soundscapes, tactile and proprioceptive displays in VR, "are generators of fictional truths, things which, by virtue of their nature or existence, make propositions fictional" (37). The display of *Microsoft Flight Simulator* makes it fictional that I am flying from Phoenix Sky Harbor to LAX in a Boeing 737; the orchestrated hemispherical projections and motion of the robotic seat in *Harry Potter and the Forbidden Journey* make it fictional that I am swooping after Harry during a game of Quidditch; the depictions in *Half Life: Alyx* make it fictional that I am getting closely acquainted with a head crab.

It is the egocentric and interactive nature of the depictions of VR that makes them so apt for such fictional uses. In their account of "self-involving interactive fictions" (SIIFs) Jon Robson and Aaron Meskin argue that a range of fictions, including videogames, role-playing games such as *Dungeons and Dragons*, and even written fictions which place the reader as a character in the action such as the *Choose Your Own Adventure* books, "are fictions that, in virtue of their interactive nature, are about those who consume them" (Robson and Meskin, 2016: 165). One of their motivations for identifying this class of fictions is "the high degree of first-person discourse that is found in talk about our interactions with them. Gamers typically make a variety of first-person claims concerning the games they are playing ('I defeated the dragon,' 'I was killed by the creeper,' and so on)" (167). To connect these ideas with the theory presented here, we might think that, in these cases, the "spectator of the picture" or "implied viewer" is given a robust identity as "I" the *player-character*. Again, the spectator is also actor in the fictive drama. And in this way, such fictions are, in a slightly different sense to how the term has been used above, *egocentric*. An egocentric picture is one where the viewer is spatially related to the scene pictured; an egocentric fiction is one where the player, viewer or reader is involved in the fiction such that their interactions generate the fictional content of the work or game in a way that self-reference in fictional terms is possible and appropriate. The egocentric pictures of VR are hence ideal in the formation of self-involving interactive fictions because the self is involved spatially, making it natural in such cases to refer to yourself and your interactions with the objects that you fictionally see *around* you. VR, in its fictive uses, comprises a spatially self-involving interactive fiction.

Connecting back to the discussion of the previous chapter, then, what is interacted with in these cases? Again, obviously the direct object of interaction is the medium, and specifically, its interface. But when we acknowledge the fictive nature of the medium in these instances, and the self-reference that occurs when we describe our interaction with it—that is, "I defeated the dragon," "I was killed by the creeper," "the head crab massaged my scalp"—the *apparent* objects of interaction in such cases are fictional things. And so, to be correct, we need to refer to the interaction with a creeper in *Minecraft VR*, or a cannibal in *Resident Evil VII* as *fictional interaction*. It is fictional interaction because it derives from a real interaction with the VR prop that makes it fictional that one interacts with creepers, cannibals, and head crabs.

Self-involving interactive fictions, particularly of the representationally rich kind seen in VR, have many uses beyond the kinds of entertainment principally associated with fiction. Some of these "serious" uses of VR, and the features of their computational media, might lead to some reluctance in acknowledging their fictional status. "Fiction," after all, might sometimes be used as a near synonym of "story" or "fantasy," and VR in many uses does not look like a story-telling device or case of fantasizing. A medical simulator, or a flight simulator used in training pilots might obviously be thought of as simulations *rather than* fictions, emphasizing a distinction that has previously been drawn by Espen Aarseth (2007). They are nevertheless fictions because they only fictionally depict their users as engaged in medical care or flight. Their simulative aspect comes about because these interactive fictions are carefully designed to fictionally realize, with some accuracy, the activities of medical care and flight. The appropriateness of their description as fictions will become even more clear when in the following sections I describe cases of VR where the technology is used to depict, and allow interaction with, non-fictional things. Nevertheless, we should acknowledge that many of the fictive uses made of VR may seem non-imaginative and very serious minded. But then this is true of fiction more generally: fictions can be used to educate, instruct, and to hypothesize, in addition to their use as entertainments.

One other use of fiction might be worth noting at this point, because it has a bearing on the earlier discussion of *trompe l'oeil* and some of the contested issues surrounding VR. Fictions can also be used to *deceive*, in the case that their nature as fiction is hidden from their audiences. *Trompe l'oeil* sometimes seems to be such a case of a pictorial deception, and we can imagine VR being used to deceive not only the eye, but also the understanding. Earlier I acknowledged in principle (though hardly in current practice) that an experience of VR might be indiscernible from reality. If such a compelling or "perfect" VR experience was presented to a user as reality, we might then question whether, for that person, the VR was a fiction. This kind of scenario is a staple in fictions such Orson Scott Card's novel *Ender's Game*, and the films *The Truman Show* and *The Matrix*, and it has recently been used to sow doubt about our knowledge of reality (Bostrom, 2003). However, deceptive fictions are not particularly mysterious, because they are a reality of our daily lives in the form of the lies that are told to us and that we ourselves tell, and I believe that these cases of VR would remain fictions even if one believed them to be depictions of reality. What matters for the fictional status of a representation is the context of intended use in which it is presented, and not necessarily how it is received. A virtual reality that continued to fool the eye even after careful inspection would remain a pictorial fiction, nonetheless.

5.3 Virtual Documentary

In addition to the fictive uses of VR, and contrasting with them, VR picturing can also be used for non-fictive purposes. I will argue here that there is more than one kind of virtual non-fiction, including those cases where virtual worlds are used to document the actual world, including *virtual documentary* specifically; where virtual technologies are used to perceive and interact with the actual and present world by functioning as a kind of virtual prothesis; the primarily aesthetic uses of VR, where the surface and its spatial configuration is the principal concern, or where the technology is used to produce artistic surfaces; and finally, the cases of mixed or augmented reality where the depiction of the actual world is overlaid with other kinds of information, visualizations, or spatial configurations. No doubt there are other cases of VR non-fiction, and more uses will arise over time as our understanding of the potential of VR develops. But these four non-fictive kinds of VR—documentary, prosthetic, aesthetic, and augmented—will structure the following discussion.

What do I mean here by "non-fiction"? We could of course simply follow the lead of the phrase itself, and contrast non-fiction with fiction: whereas fiction deals with the imagination and imaginary things, non-fiction does not involve make-believe or the imagination but concerns what is real or actual. This would be problematic, however, because the imagination suffuses our dealings with this, the actual world, and is hardly restricted to the enjoyment of novels, films, videogames, or fantasies. Moreover, works of non-fiction, and even factual documentary—surely a central case of non-fiction—frequently involve fictive aspects or the employment of fictive techniques: the "reconstruction" or "docudrama" is a frequent feature of documentary film and factual television programming, for example.[3] Equally, contrasting fiction with "fact" would also bring with it complications, not the least of which is that many fictions are filled with facts, and factual works filled with fictions or falsities. We should not expect a clear separation between fiction and non-fiction or factual works in their instances. Derek Matravers has expressed some skepticism about the distinction between fiction and non-fiction itself, and specifically, doubts about invoking the role of the imagination to distinguish the two (2014). I have a certain amount of sympathy for his views; however, it is not my task to solve the big issues of philosophical aesthetics, but rather to show that VR can profitably be considered to be a new player in them. Here I will persist with the terms fiction, non-fiction, and factual, because they allow us to identify a range of unexpected cases of VR that permit us to escape the image of VR as always involving the depiction and interaction with shadowy virtual worlds that are not quite real, or are merely conceptual or imaginary. There is, I think, a perfectly evident difference between cases of VR that depict imaginary or fantastical worlds and uses of the medium that aim for an engagement with a sometimes quotidian reality.

Perhaps an example of VR used for factual purposes will be the best way forward. A good example that makes the relevant distinction is Google Earth VR. Google Earth VR uses the assets developed for Google Earth—3D imagery drawn from satellite and aerial photography, in addition to Google Street View—so that they can be viewed on the Oculus Rift and HTC Vive stereoscopic viewers. The user can navigate through this imagery using the VR controllers to explore the world. In terms of the theory developed here, Google Earth VR comprises an egocentric depiction in which the intentional object is the Earth's geography, such that one might inspect and learn about it. This exploration is no more fictional than viewing a globe or atlas, and in many respects provides a more faithful depiction of the world's features than the flat surface of a map.

Let us pause though to consider an alternative view of this example. When I view a Google Earth VR depiction of New York City, we might be tempted to think that I am fictionally observing New York, because after all, I am not actually looking out over the city. And, as above, it is very easy in this case to frame the use of the app as one of *exploring the world*, which seems to encroach on fictionality. However, once again this is misdrawing the division of labor between perception and imagination, and unnecessarily taking the former to amount to the latter. When I watch a television documentary on New York City, it is not fictional that I am seeing New York, or that I am exploring the city. Rather, it is simply that I am watching a pictorial documentation of New York City, and I can do so because perceptually I am able to reconstruct the visual markers of the surface into the spatial configuration of the picture, and then to see this as a video documenting New York City. At no stage in this process do we need to posit anything fictional. And if we seem to do so—as with the phrase "explore the city"—this is likely a colorful way to refer to the act of viewing pictures. Of course, a slightly different case might involve fiction, and it is illustrative: earlier in its development there was a flight simulator in some versions of Google Earth, in that one could fly an aircraft around the depicted geographical areas. In this case it does look like there is a fiction involved because in the addition of the activity of flight it looks very much like there is a fictional spectator of the pictured scene who is the looking down on New York City from their vantage point in an aircraft or exploring the city by flying around it. But this, in essence, is equivalent to the fictional flying available in *Microsoft Flight Simulator*, which now also uses ground satellite data in the creation of its depictive assets.

Another form of factual VR is VR documentary specifically. In this case, the technology of VR is used to produce documentary features similar to those encountered in film or television, though with differences due to the egocentric and interactive nature of VR. One example is *Travelling while Black*, a documentary directed by filmmaker Roger Ross Williams, which is available for viewing on the Oculus Quest stereoscopic headset. It details the challenging experiences of African Americans traveling in the Jim Crow era. The film is displayed in 360-degree video with stereo sound, and includes interviews and dramatizations relating to the historical events. Such VR documentaries often do involve aspects of fiction in their dramatization, as is common with film and television documentary, but the factual subject—in this case racial discrimination in America—distinguishes these cases from the fictional uses of VR explored in the last section and aligns them more closely with traditional forms of documentary making.

But again, we may question whether VR is properly placed within the established category of documentary, perhaps thinking the differences that VR bears to film and television documentaries motivate describing these as simulations rather than documentaries. So, what is documentary itself? For the philosopher of film Carl Plantinga,

> When a filmmaker presents a film as a documentary, he or she not only intends that the audience come to form certain beliefs, but also implicitly asserts something about the use of the medium itself—that the use of motion pictures and recorded sounds offer an audiovisual array that communicates some phenomenological aspect of the subject, from which the spectator might reasonably be expected to form a sense of that phenomenological aspect and/or form true beliefs about that subject.
>
> (Plantinga, 2005: 111)

VR documentary, despite the difference in its medium, fits with this analysis. The difference in VR is in how these goals are achieved, but also, perhaps, in how the "phenomenological aspect" of VR—an egocentric picture—might alter the phenomenology or even the epistemic warrant inherent in the medium. Being "within" the picture may change how users derive true beliefs about the subject of the documentary.

This leads to the question of whether VR documentary in fact has such advantages over traditional forms of the genre. It could be argued that the presence or immediacy of virtual documentary gives viewers a more impactful sense of the reality of the issues covered. Being virtually located within Ben's Chili Bowl, a Washington DC restaurant that catered to African Americans during the period of segregation, may give the events a sense of immediacy to the viewer. Media and communication theorist Kate Nash notes that,

> The most significant difference between VR and the audio-visual practices (including those in the realm of interactive documentary) which have traditionally preoccupied documentary scholars, is that in the case of VR our practices for depicting the real are no longer contained in the same way by the apparatus and aesthetics of the (2D) screen. While the screen has not disappeared in an ontological sense, at the heart of VR is the production of the illusion that we have entered into and become a part of the world that we used to watch on the screen.
>
> (Nash, 2018: 97)

The implication here is that because VR no longer relies on the screen, but places the viewer within the scene, the viewer may have a more direct (though still illusory) access to the documented situation. I doubt however that the genuine media differences of VR make it especially epistemically merited, as the kind of interactive egocentric picturing I have described here is just as prone to the doubts about the inherent truthfulness of documentaries, as any other kinds of pictures are. *Travelling while Black* surely presents an insight into the issues it depicts, but that it presents a more direct experience of Jim Crow era racial discrimination itself, because of its virtual medium, would be a claim fraught with complications. VR documentary, while it might benefit from some of the experiential features of VR, most obviously the experience of presence, is likely to raise many of the same issues that affect other forms of documentary, which may present myths, propaganda, and lies, even if the form is intended to be veridical (Plantinga, 2005: 113–114). And in computational media such as VR those problems may be more worrisome: because of the frequent reliance of VR on CGI, and with the seemingly increasing epistemic untrustworthiness of the latter medium owing to digital photo manipulation and "deepfake" animations, the ability of VR to depict reality such that we can form true beliefs from its encounter, might be compromised. I cannot explore the connections between VR and epistemology in this book, though I will briefly return to the issue in the next section.

Nevertheless, if in its use of egocentric picturing VR documentary is no better off epistemically when judged in terms of its truthfulness, this mode of viewing might be in at least one way more impactful. This is because VR might document the experience of places or events otherwise inaccessible to the viewer, but feasible as a case of VR picturing. *Memoria: Stories of La Garma* is a narrated VR experience allowing for the virtual exploration of the La Garma cave complex in Cantabria, Spain. Initially installed in the Museum of Prehistory and Archaeology of Cantabria in Santander, nearby the caves themselves, users may virtually explore the recently discovered caves and their array of artefacts and paintings. These caves are ordinarily inaccessible to viewers due to the prehistorical sensitivity of the area, but when mapped by laser scanning and photogrammetry and depicted in VR, many people can have the experience of viewing the La Garma complex without damaging the actual caves. Note how this is a modern technological parallel of the physical reconstructions that were sometimes built in the proximity of caves, including those in Cantabria and also the more famous caves in Lascaux, showing that VR may have the potential to replace other forms of documentation.

Other VR experiences document places inaccessible to us because they are in the past. I mentioned at the beginning of the book a VR documentation of Ancient Rome, but this is in fact a very popular destination for VR documentation, and one can find multiple virtual reconstructions of the ancient city. Other VR experiences may virtually document the more distant past, by depicting ancient ecologies and life forms as in *David Attenborough's First Life* (2017). This experience, initially shown in London's Natural History Museum, depicts life in the Cambrian oceans of 540 million years ago. Similar uses of VR can now be found in museums around the world, allowing users to experience art, history, and science, and indeed, are perceived by museum professionals as both an opportunity, and challenge to the very idea of the modern museum because of how they remediate its experience (Carrozzino and Bergamasco, 2010; Shehade and Stylianou-Lambert, 2020).

Other VR experiences document objects, places, or experiences that due to technical constraints such as danger, costs, impracticality, or inefficiency, are more effective if rendered for their spectation in VR. In these cases, the VR may often go proxy for the real experience, so that one can learn or prepare for it, such as in VR experiences that allow surgeons to prepare for complex spinal surgeries (Croci, et al., 2020); the depiction of virtual animal anatomy for the purpose of veterinary training (DeBose, 2020); or to document the bodily movements of athletes to understand "the perception-action loop," that is, "how perception influences choices about which action to perform, and how those choices influence subsequent perception" (Bideau, et al., 2009).

Some VR experiences may even document spaces that would be literally impossible for an actual viewer to experience. In this way one may explore the cosmos in *Overview VR* (2018) by rapidly moving though it in a way that no perceptual agent could, and via manipulating and rescaling the depiction to get an impression of the true vastness of space. Similarly, at the other end of the scale of size, one might use VR to view and interact with dynamic molecules, an interaction which recent research found "improved understanding and insight into complex molecular systems, furnishing an improved sense for how molecular objects move and respond to perturbation, facilitating efficient clear communication, and encouraging researchers to think creatively about their systems" (O'Conner, et al., 2019). Hence in some cases at least, the apparent virtual perceptual acts fostered by egocentric picturing might allow their users to conceive or understand things more easily than when encountered in other representational media.

5.4 Virtual Transparency

If we do consider that some cases of virtual reality depict an *interaction with fictional worlds, objects or characters*, and that other uses *document the actual world*, by mixing the interaction of the first kind of VR, with the factuality of the latter, a potential new use of VR media comes into view: namely, the use of VR to remediate *an interaction with real items or the real world*. And there are now such cases where VR is employed for such real-world interaction. One such case is the Disney research on ball catching in virtual worlds referred to at the beginning of this chapter. Here, credibly, the spectator in the picture might be identified with the actual user of the VR headset, and not only do they see the egocentric picture, but they see through this to perceive items in the actual world around them, specifically, the ball that they catch. This experiment shows that in VR, not only is the egocentric perspective tracked, but items from the actual world can also be tracked and placed into this egocentric space such that they can be the object of real interaction. If this is the case, then perhaps the user *literally sees through the VR picture to the item depicted*. The VR user really sees the ball, because this, after all, is what allows them to catch it.

To understand how this might be the case, I need to introduce a debate that has been conducted within aesthetics over the last forty years regarding the epistemic features of various forms of picturing. In "Transparent Pictures: On the Nature of Photographic Transparency," Kendall Walton presents the claim that "the invention of the camera gave us not just a new method of making pictures and not just pictures of a new kind: it gave us a new way of seeing" (1984: 251). The development of photography is not the only technological extension of seeing, in that,

> [m]irrors are aids to vision, allowing us to see things in circumstances in which we would not otherwise be able to; with their help we can see around corners. Telescopes and microscopes extend our visual powers in other ways, enabling us to see things that are too far away or too small to be seen with the naked eye.
>
> (251)

Photographs, according to Walton, are transparent in that we literally "see the world though them." Specifically, they enable us to "see into the past" (1984: 251). I have here argued that VR media, and their nature as egocentric pictures, are a further technological extension of picturing. Do they too extend what we are able to see through them? The answer I believe is yes;

(some) egocentric pictures are transparent in that they allow their users to see (and potentially interact with) items that need not be in their actual presence. I will refer to this thesis as *virtual transparency*, and it is an idea I have defended at greater length elsewhere (Tavinor, 2019a).

Why does Walton think that photographs are transparent? He argues that for a medium to be transparent, such as a mirror or telescope is, there are constraints on "how we acquire information" from the medium (1984: 262). Contrasting photography with traditional paintings, the latter of which he argues are not transparent, in photographs, what ends up on the picture surface—and hence, what is represented—depends on the physical mechanism involved in the photographic process; whereas in painting, the intentions of the artist also partially determine what is depicted on the surface. Imagine a landscape painting and a photograph, both depicting New Zealand's highest mountain, Aoraki Mount Cook. That both the painting and the photograph have the features they do is counterfactually determined by the nature of the scene itself: if the appearance of the mountain had been different—if, say, it had not been snow-capped—the pictures themselves would be different. But Walton asks: "why are these counterfactuals true?" (264). In the case of the painting, it is because the painter's beliefs about the mountain would have differed between the cases (if they were appropriately acquainted with facts about the mountain). However, in the photograph, the counterfactual is true because the mechanical process involved in the production of the picture could not help but capture the difference. Hence, photographic images are *non-intentionally mediated*, in that such images "are counterfactually dependent on the photographed scene even if the beliefs (and other intentional attitudes) of the photographer are held fixed" (264). According to Walton, this is why we typically ascribe to photographs a special realism or epistemic merit beyond that of painting, and why we really do see through them to the objects they depict.

Because they are non-intentionally mediated, some cases of VR depiction meet these conditions of transparency: what is depicted in Pan and Niemeyer's VR experiment is counterfactually dependent on the scene itself, but in this case, it is not a mechanical process that assures the counterfactual, rather an *algorithmic* one. The position and movement of the ball depicted in the VR space is counterfactually dependent on the position of the actual ball because it is tracked and depicted in 3D space by the algorithms at the heart of the VR system. Walton, of course does not mention VR in his paper, but one of the hypothetical examples he discusses suggests he may not be opposed to the conclusion drawn here. He introduces Helen, a woman whose optic nerves are connected to a device that allows a doctor to stimulate them to receive images

> corresponding to what he sees, with the result that she has "visual" experiences like the ones she would have normally if she were using her own eyes… Helen *seems* to be seeing things, and her visual experiences are caused by the things she seems to see. But she doesn't really see

them. [...] It is only because differences in scenes make for differences in the doctor's beliefs that they make for differences in her visual experiences.

(Walton, 1984: 265)

But Walton thinks a second hypothetical example does describe a case of transparency, and the situation he describes lines up in many respects with VR seeing:

> Contrast a patient who receives a double eye transplant or a patient who is fitted with artificial prosthetic eyes. This patient *does* see. He is not relying in the relevant manner on anyone's beliefs about the things he sees, although his visual experiences do depend on the surgeon and on the donor of the transplanted eyes or the manufacturer of the prosthetic ones.

(265)

Similarly, in the VR case, the VR peripherals and the algorithms that allow them to track and produce the images of the moving ball are built and programmed by technologists. But once they are set up to achieve egocentric picturing, what is depicted is determined by the tracked scene itself. Indeed, the Pan and Niemeyer experiment produces a kind of prosthetic vision akin to Helen's prosthetic eyes, with the *Rube-Goldbergian* complication that the viewer also has functional eyes of their own (Tavinor, 2019a: 146).

Walton's photographic transparency thesis has been controversial, and one critique of the view allows us to see how virtual transparency may even be more credible than Walton's claims about photography. To really see something through a medium, claim the philosophers Jonathan Cohen and Aaron Meskin, rather than merely having a visual experience caused in a non-intentionally mediated way, "what is essential is that the relevant visual experience is produced by a process that carries egocentric spatial information about the object" (Cohen and Meskin, 2004: 198). Cohen and Meskin base this judgment on a more general account of what it is to see a given object, arguing that "x sees y through a visual process z only if z carries information about the egocentric location of y with respect to x" (210). This is why they take it that we can see objects in mirrors—which do convey egocentric spatial information bearing this counterfactual relationship, even if the precise spatial relationship cannot be judged by the viewer, such as in a carnival "house of mirrors"—while we do not see things in photographs, which are indeterminate with respect to the egocentric spatial position of the items they reveal. They note that

> the visual process of looking at photographs fails to carry egocentric spatial information about their depicta. For there is no probabilistic relationship between the photographic image and the egocentric loca-tion of the depictum: as I move around the world with the photograph,

the egocentric location of the depictum changes, but the photographic image does not.

(Cohen and Meskin, 2004: 8–9)

When you hold a photograph of the Taj Mahal before you, and take two steps to the right, the egocentric location of the Taj Mahal changes, but the surface of the picture does not change to reflect this spatial orientation. Cohen and Meskin argue that this is enough to show that looking at the picture is not equivalent to looking at the Taj Mahal itself.

But this egocentric spatial information is precisely what I have argued distinguishes VR picturing from other pictorial forms. When I move around the room while wearing the VR headset, the egocentric location of the scene and objects depicted in the pictorial surface change in concert with this movement. And it may be because VR, at least in non-fictive uses, can be used to provide the viewer with egocentric spatial information about objects, that its users do see the things depicted in such cases (or at the least that VR pictures escape the objection that Cohen and Meskin frame against the thesis of photographic transparency). Moreover, there is a positive reason to think that VR in such cases is transparent. It is precisely because the depictions on the headset in Pan and Niemeyer's experiment do bear an egocentric counterfactual spatial relationship to the ball that the participant is able to catch that ball. What could be a better explanation of this than that the user *sees the ball through the depiction of a VR headset*?

Although there is more to Cohen and Meskin's argument against photographic transparency, and to my own possible defense of VR from this argument, it can be concluded that VR is in one respect at least, more credibly seen as a case of transparency than photography is. In the ball-catching experiment, the virtual medium preserves the function of seeing (and catching) a real ball, partly by the provision to the user of the kind of information that some see as constituting one of the characteristic or even definitional features of actually seeing objects. Moreover, I take it that what is occurring in this case of VR picturing is that the user can interact with the depicted object because it has a real and present existence for the user, even though the perception of its presence is mediated through a VR headset. The consideration of the "intentional object" of pictorial seeing is again useful here. Two users of different VR picturing applications might perceive the same pictorial surface, in that they may perceive identical 3D configurations of a ball that is moving toward them in their perceptual space. But despite this configurational identity, the cases may differ with respect to the existence of the ball depicted by the egocentric picture. The ball may really exist, and thus really be seen and interacted with, or it may be *merely apparent* because it is depicted as part of a fiction (say, if the ball catching was part of a VR videogame). This, essentially, is a difference in the intentional pictorial object, in that in the first case VR is in service of a behavioral response to the actual present world, and in the second case it serves the imagination. Hence, we can, and sometimes do, perform real and not merely fictional actions on virtually depicted objects. VR transparency, is, following the analysis of virtuality provided earlier, a non-customary remediation of the functional efficiency of ordinary perception and agency.

This account of virtual transparency also has an effect on the appropriateness of various terminology regarding VR. In some of the literature, "VR fictionalism" is taken to be the claim that the objects depicted in VR are fictional, and it is a position that has been attributed to my own work (Chalmers, 2017). Correspondingly, one understanding of "VR realism" is that the objects interacted with in VR are real and not fictional, even if at the same time they are virtual, socially constructed, or the like (Brey, 2014). But under my view, "VR fictionalism" refers to those undeniable cases where VR depicts fictional items; "VR documentary" refers to those cases where VR documents non-fictional objects or places; and "VR transparency" refers to, perhaps as a special instance of non-fiction VR, those cases where VR depicts real items that are *virtually present* to the user. I will retain the term "VR realism," but stripped of its ontological implications, to refer to the issue of whether the medium is especially realistic in how it *appears* to users.

In addition to allowing a user to catch a ball, there are other cases of VR picturing that can be understood as cases of the remediation of perception and agency afforded by VR transparency. The use of VR haptics may allow surgeons to really feel tactile feedback while performing robotic surgery training, to "reduce surgical errors and potentially increase patient safety" (Van der Meijden and Schijven, 2009). Virtual reality drone flying is another case, where a user may fly a drone equipped with cameras to allow for VR spectation as they fly. Perhaps most compellingly, the car manufacturer Toyota is developing robotics systems to allow for the VR spectation and control of a robot's movement. The T-HR3 humanoid robot currently uses HTC Vive for this purpose. In a Toyota corporate press release, the idea of transparency and remote VR control is explicit when they refer to "'*virtual movement,*' when an operator's body or body part is virtually moved through a remote space via avatars or agents" (Toyota, 2019). Many of these technologies are still in their infancy, and sometimes the use of "VR" to describe these technological developments owes more to marketing than to an accurate description of the technologies themselves. But we easily can imagine that soon VR transparency will have many applications, potentially allowing for prosthetic seeing and interaction with remote or inaccessible objects and places, and, as the designers of the T-HR3 envisage, an increased sense of virtual "mobility."

In most of these cases of VR transparency, unlike the ball catching experiment which indeed is atypical, the act of viewing is *remote* from the objects perceived, in that the egocentric perspective of the picture does not align with that of the actual viewer but is located in the position of the prosthetic appendage, drone, or robot actually present to the objects seen. This is perhaps the standard case in VR prosthetic seeing because the utility in the technique is seeing what one cannot ordinarily see, perhaps because it is at a distance or otherwise inaccessible. But is this a problem for the claim that the user sees the objects depicted? In some of these cases it seems like the egocentric criterion of seeing held by Cohen and Meskin breaks down: the apparent perspective of the drone or robot is in a counterfactually dependent spatial relationship with the objects seen, but the actual viewer may not be. Here is an example that shows why. Imagine that the person in control of a VR military drone is travelling in an aircraft circling the area where the VR drone is itself really flying (doing whatever VR military drones might do in such cases). In this case, one aspect of the movement of the tracked perspective of the user through space—its movement within the aircraft circling the theatre of action—has no effect on the perspective depicted to the user, and hence on the egocentric relationship they bear to the depicted objects. It is not just that the viewer may not know the location of the objects depicted—in reference to their own position, or absolutely— even as they interact with them, but that there is no such dependent relationship.

In this case VR seems much more like a photograph than a customary case of seeing, but the drone pilot is still able to interact with the scenes and objects that are their real interest, and it is my intuition, at least, that they still see these things to do so. This seems like a problem either for my claim that VR is a case of remote seeing, or for Cohen and Meskin's claim that genuine seeing requires a counterfactually dependent spatial relationship. The basic question becomes, does VR allow us to remotely view objects and scenes? It is my view that this is the case, though admittedly largely based on my intuitions about these examples. And this is not so far removed from how we think about other cases of remote listening and viewing, such as telephones and TV broadcasts, where we may without much confusion say that we heard or saw something from which we were spatially dislocated. As I have concluded elsewhere, it may be that Cohen and Meskin's theory is too conservative in what it counts as genuine seeing of objects (Tavinor, 2019a: 154). We may well fall back to Walton's position and conclude that, not only is VR transparent (because even photographic images are), but it also adds the elements of spatiality and interactivity to this transparency, allowing for a new way of seeing and acting.

Thus, if this is prosthetic seeing, then relevant here is another of the observations about virtuality made earlier in this book: that, after a period of social familiarization, virtual items so easily come to be regarded as actual instances of the items they instantiate in a virtual way. By preserving the crucial function of stores—allowing for the sale and distribution of commercial items—virtual stores can be regarded as stores *simpliciter*. A conversation on a telephone is not a virtual conversation, but just a conversation. The preservation of function allows these virtual items to instantiate items of their given kinds, despite the change of medium. Transparent pictorial seeing can be considered another such case, and so its consideration clearly cuts to the heart of the nature of virtual media, where it seems to be a structural or functional correspondence between the actual item and its virtual counterpart that comprises the fundamental feature of virtuality. In the cases of transparency above, the function and structural correspondence is in terms of the remediation of experience and agency in the real world, and as such this is perhaps the clearest case of VR fitting the definition of virtuality provided in Chapter Two of this book.

5.5 Augmented and Mixed Reality

This is a convenient time to discuss the related phenomenon of augmented reality (AR). Augmented reality involves the representations of objects and scenes of the actual world being supplemented by other graphical elements and representational devices. The typical cases are visual, where images of an actual scene may be overlaid by additional graphical elements, either providing information about the scene, or augmenting its content in various ways. These elements are usually spatially indexed to a location in the real scene, giving the impression that they occupy space in the scene. So, for example, the famous AR game *Pokémon Go*, that preoccupied many of us for a month or so in 2016, depicts Pokémon as inhabiting the actual world and placed alongside familiar places and landmarks such that one can seek them out.

Where might augmented reality fit in to the analysis of VR presented here? One established view is to see augmented reality as a "subset" of VR (Milgram and Kishino, 1994). Given the vagaries of the usage of the term "virtual reality," there is no great reason to disagree with this; however, in terms of the specific theory of VR developed here, we can also distinguish between those cases of AR that employ egocentric picturing and those that do not. Most cases of AR, it should be noted, do not involve VR in the sense described here, because they ordinarily use 2D picturing surfaces and 2D pictorial augmentations. *Pokémon Go* is depicted on the 2D surface of a phone or tablet, and the pictorial augmentations are 2D rendered Pokémon, which give only a basic impression of spatiality by utilizing 2D pictorial conventions. Two-dimensional AR has many uses though and raises in its own way many of the same issues discussed regarding computer media and picturing, tracked vantage points, interactivity, and so on.

In the sense of VR media developed here, VR augmented reality does exist, however, and moves have been made in the close combination of the two technologies, so that one can encounter depicted objects placed within actual egocentric space. The combination of VR and AR is sometimes unhelpfully called "mixed reality." This designation is undesirable because the phrase gives the unwanted implication that the mix is *ontological*, as though reality, augmented reality, mixed reality, and virtual reality sit on an ontological continuum, from the more to less real. This kind of *graded* account of the ontology of these media is unfortunately common and I will have something to say about it in the final chapter. Nevertheless, mixed reality involves the projection of 3D spatial objects into an actual environment via a stereoscopic viewer, so that visual augmentations might seem to occupy a space in the actual scene as objects there. It is in this sense that

Microsoft's HoloLens supports mixed reality, and its apps provide good examples of the medium. *HoloAnatomy* (2016) is an app published by Case Western Reserve University, that allows the user to inspect elements of human anatomy. The pertinent difference between VR augmented reality and basic 2D augmented reality, is that VR more effectively places the graphical and informational elements into the viewer's spatial situation because of its involvement of egocentric picturing. The items are not only indexed to the space but are so in such a way that they are spatially congruent with the user's own space, and they allow for "viewing in the round" as the viewer moves around them, and even the potential for interaction of the kind discussed in the previous chapter.

What is the status of these AR depictive forms, and how do they fit into the categories of fiction, documentation and transparency introduced here? It is sometimes claimed that AR augments reality, whereas VR replaces or excludes it. But as can be seen above in the case of virtual documentary and transparency, this is not the case, as many instances of VR seek to reveal reality in ways in which we cannot presently experience it. Rather than such blanket statements, it should come as no surprise, that whether AR presents fictional, documentary, or transparent elements will very much depend on the context of use. *HoloAnatomy* is a case of mixed reality documentary: beyond its use of egocentric picturing and augmented reality, in form it is basically equivalent to the anatomy books or anatomical models also used by students. A case of transparent mixed reality comes from the Disney research: in a version of their experiment, Pan and Niemeyer also depicted a predicted trajectory, and noted that "[t]he predictive assistance visualization effectively increases the user's senses but can also alter the user's strategy in catching" (2017: 269). Here, in VR, the egocentric depiction includes information about the location at which the ball could be intercepted, and users change their behavior to utilize this information. And, of course, there are more fantastical uses of mixed reality technology, where it can be used to place fictional creatures or objects within the depicted environment, and even allow for fictional interaction with such things. *Dr Grordbort's Invaders* (2018), a game developed by Weta Workshop for the Magic Leap mixed reality headset, involves defending your living room—or wherever you happen to be playing the game—from invading robots.

There are many other documentary, transparent, or what might be more generally called "serious"—that is, non-entertainment—uses of VR media beyond those detailed here, and no doubt many more to be discovered and developed. Here I want to note just one more widespread serious application of VR that does not sit squarely in the categories identified above, but which is relevant here because of its underlying assumptions about the nature of VR. This wide range of cases exists where VR is used as a medical or psychological treatment, therapy, or rehabilitation. Here, it is not the fictive or documentary information inherent in the VR image, or the uses of VR picturing to interact with the actual world, that is the focus, but how the VR experience might itself be of therapeutic use because of the kinds of remediated experiences it makes available in the controlled and safe environment of VR. This potential use of VR is enormously widespread, including virtual exposure therapy for the treatment of anxiety (Anderson, et al., 2013) and phobias, including "fear of flying, spiders, and heights, as well as panic disorder, agoraphobia, social anxiety, and posttraumatic stress disorder" and blood injection (Jiang, Upton and Newby, 2020: 637), as a neurorehabilitation tool for brain injuries (Aulisio, Han and Glueck, 2020), and providing rehabilitation for Parkinson's disease (Dockx, et al., 2016) or stroke victims (Laver, et al., 2017). The sheer breadth of such applications and their speculative nature might make one skeptical that some of these applications amount to a virtual clutching at straws. This is not the venue to express this skepticism fully, but it is worth noting, because these aspirations for the therapeutic uses of VR illustrate, first, the assumption or intuition that VR in fact does remediate perceptual and bodily experience of reality in a way that would suggest it for therapeutical uses; and second, the hopes (realistic or not) that VR will positively transform many different domains of human life.

5.6 Virtual Aesthetics and Art

In a book on virtual aesthetics, it would be natural to hear something about art. VR has obvious artistic uses, and artists have already engaged the medium in fascinating ways, and these may encompass all the categories previously considered in this chapter. Given the uses outlined above—fiction, documentation, and transparency—it is already easy to see that VR has potential as an artistic medium because of how these features themselves have established artistic functions. Fiction and documentation have obvious artistic purposes, and perhaps the most obvious artistic use of VR is in cinema and documentary, of which I have already introduced examples above. I also suspect that art may employ VR media in a further way, by directing our attention to the medium itself, in a way that art has frequently done. Quite differently, VR might act as a medium for performance.

In cases such as VR documentary and cinema, while the object of appreciation is a VR object, the *production* of the object of appreciation may not itself involve perceiving and interacting *within* a VR medium. But other cases allow for VR media themselves to become a medium for creation, so that one is situated in VR during the creative act. Perhaps the most obvious and popular such case is the VR painting application, Tilt Brush. Formerly of Google, but now an open-source application, Tilt Brush allows artists to create 3D paintings in egocentric pictorial space. Using the various virtual brushes, the user marks the surface to produce the image "around them," and many artists express surprise at the experience of producing a painting from "within" it. Indeed, it may be more accurate to refer to these creations as virtual sculptures than paintings, given their apparent three-dimensionality. The British portrait artist Jonathan Yeo has employed this potential of

Tilt Brush to produce a virtual self-portrait that was then realized as a sculpture in bronze. The production of the sculpture engaged with these virtual technologies at several steps: Yeo's facial profile was 3D scanned and then imported into Tilt Brush, and working from this model, Yeo used Tilt Brush to virtually sculpt the portrait. This graphical object was then 3D printed, with molds being made from these parts, allowing the sculpture to be then cast in bronze.

VR may even count in some instances as a medium for performance. A performing art is one in which the actions of a performer are the focus of appreciation for an audience, unlike arts such as painting and film where an object is produced, and it is this object that is the focus of appreciation. Virtual reality has the artistic benefit of allowing audiences remote from the performance to view and appreciate the performative act via a VR headset. What kinds of performance might be possible in virtual spaces? Dance is an obvious activity that can be tracked in virtual depictive spaces, and indeed, virtual reality dance already has a reasonably long history (Smith, 2018). Because it is possible for the user's bodily movements and gestures to be tracked into a virtual space in "room-scale" VR, viewing of live dance performance in VR seems a possibility. Nevertheless, recent examples of dance in virtual media are mostly confined to prerendered 360-degree videos, such as in *Lifeboat* (2017), a short music video starring the dancer and singer Shannon Rugani, directed by Luke Willis. Virtual media also allow for the augmentation of dance performances, either through the overlaying of the performance with additional graphical elements, or the allowance for audience interactivity. One example is the VR film *Anicca* by American animator Cecilia Sweet-Coll, where motion-captured dance is used to produce often-abstract 3D animations. Other performance arts are also possible, including virtual musical instruments where the user's gestures may be tracked in a way that allows for interaction with a VR depicted instrument.[4] Hence, virtual technologies make available this kind of performative avenue, and they may also bring these performances to remote audiences in a way that conventional and non-interactive media cannot.

The creation of artworks or artistic performances might utilize VR, but what, if anything, might VR depiction add to *aesthetic appreciation*? That VR need not involve aesthetic attention to the picture surface is a very tempting conclusion to make given what we know about the typically non-reflective nature of VR pictorial seeing (that is, that users typically have no aesthetic concern with the surface at all). This raises the question of whether the pictorial surface in VR *ever* factors into the aesthetic appreciation of the apps, games or works of digital art that employ VR, and whether its aesthetic features are ever the focus of attention. One benefit of the twofold theory of seeing is that it allows us to frame how viewers see the aesthetic qualities of the surface even as they perceive the object depicted in the picture. So, they may see the apple depicted in a still life, but also see the brush work that lends the apple depictions a realistic impression, and these two facets of the pictorial seeing may combine in the painting's aesthetic experience and appreciation. It is an interesting question whether seeing the surface factors into our aesthetic appreciation of VR media, and if we aesthetically appreciate how the spatial or depicted content of the VR world is embodied in the surface. Viewers of VR must perceive the elements of the surface, because it is through these that they perceive the spatial configuration represented by the surface, but this is consistent with them not paying the surface any great attention or even recognizing the surface as a surface, and hence not perceiving its aesthetic qualities. We also found earlier that the pictorial surface in VR is complex, being a combination of two images, viewed through lenses, and that these fuse together. Viewers do not typically perceive the two images as not fused, and so perhaps there are elements of the surface that viewers are not aware of. But are there elements of the VR surface that are seen and appreciated for their aesthetic qualities?

I have a rather lowbrow example that might seem a bit out of place in a section on VR and art, but which nevertheless illustrates how the depictive surface is sometimes prominent in VR pictorial media. In *Minecraft VR*, the VR experience can be had in two ways: first, from the perspective of being in the world of *Minecraft* itself, and second, in what is called "Living room mode." In the latter, one is placed within a VR living room, depicted in the style of *Minecraft*, in which one can play 2D *Minecraft* on a screen within the living room. It is an odd, but fun, design choice, comprising a kind of reflexive joke in which VR *Minecraft* amounts to merely playing the 2D game in a VR setting. But it is also one that brings attention to the qualities of the surface, and how they give rise to the spatial impression of the room. The surface is prominent because it is stylized in a way drawn from (but not precisely replicating) the blocky style of *Minecraft* itself. And in this case, even if there are aspects of the surface that escape the viewer's attention, there remains much that does not, and viewing these pictorial elements is no more difficult than it is in conventional pictures. The surface here is prominent because it is stylized in a way that draws attention away from what is depicted. As such, *Minecraft VR* is an example from gaming that gives credible evidence that users can aesthetically appreciate the pictorial surface of VR. Other VR art, probably more obscure than *Minecraft*, may have much the same effect.

To sum up then, in actual practice there are two folds in the perception of a VR picture, and in addition to this the perception of an intentional object, and how these present in an individual case is determined by the specific use of the medium. The first object of perception is the surface, comprising the screen or screens seen through the binocular apparatus that allows for stereoscopic viewing of the offset facets of the screen(s). This is clearly much more complex a surface than that found in other picturing forms. The second fold is the intuited 3D configuration that our visual systems perceive in this screen, and which appears for the user in their apparent egocentric space. The images on the screen are themselves produced such that stereopsis, and the allowance of the user's apparent position to be tracked, produces the appearance of coherent and stable spatial configuration. Finally, the intentional object comprises the apparent object depicted, but again, this is complex, with at least four kinds of intentional relationship being possible. First is where what is depicted is a pure fiction; second is where VR is used to document some real item that is not present to the user; third is where VR depicts an object really present; finally, and somewhat atypically, is where the question of reference does not arise, but where it is the apparent spatial configuration itself that is the focus of the VR application and its appreciation. Mixed reality may involve all of these, potentially simultaneously.

What can we conclude from the *variety* of implementations of the VR remediation of egocentric interactive space explored in this chapter? For one, having seen the practicality of VR and its frequent documentation of, and connection to, the actual world, we might be much less tempted to see VR as merely "conceptual" or "illusory," or as an attempt to "exclude" the real world. VR is grounded in, and connected to, both the facts of the actual world, and our epistemological and practical interests there, even if it sometimes allows for flights of fantasy from the world. This undeniable variety in the potential uses of VR also illustrates a technical feature of VR in that the origin and method of construction of the 3D models at the heart of VR pictures varies considerably and is usually determined by the specific use. The images themselves may borrow from other media, as in the satellite and aerial photography of Google Earth VR; in *Memoria: Stories of La Garma*, 3D scanning of real objects may begin the process; or, as in the paintings produced by Tilt Brush artists, creative gestures may be essential to the depictive process that lays down the marks on the VR surface. It may also be that many of these uses have obvious aspirations to capture their subjects in a realistic manner, utilizing the potential of VR to do so. The next chapter takes up realism as its central concern.

Notes

1 A demonstration of the technology can be seen here: https://www.youtube.com/watch?v=Qxu_y8ABajQ
2 I am a little hesitant to use the term "intentional object" here. The concept has a long contentious history and encounters problematic cases such as fictions where an intentional object might seem to be non-existent. However, the term is also useful because it does identify a relationship that pictures often bear to "external" objects, and this picturing relationship is crucial in much of what follows. I do not think that it is my job to resolve the problems with the apparent intentional objects of pictures, as this is an issue that VR shares with other kinds of picturing and non-picturing representational media. Tim Crane (2013) develops a theory of representation-dependant intentional objects that might be useful in the current context.
3 The very idea of documentary, and the possibility of its definition, or at least characterization, has led to a significant chunk of philosophical literature, usefully documented by Carl Plantinga (2005).
4 A 2019 "live" demonstration of the Microsoft *Hololens 2* mixed reality headset teased a holographic piano, though it is hard not to be skeptical about the staging of the demonstration. https://www.youtube.com/watch?v=uIHPPtPBgHk

6 On Virtual Realism

6.1 The Plausibility of Virtual Realism

There is an initial plausibility to the idea that the depicted worlds of virtual reality are, in some sense at least, more *realistic* than worlds depicted in other artistic media. Take for example *Resident Evil VII: Biohazard*, the survival horror game I introduced at the beginning of this book. One can also play the game by viewing a conventional 2D screen, but in my experience, playing through the game in its VR mode is the more perceptually striking and disconcerting experience. The sense of being within the world, the realistic appearance of the environments, and the feeling of anxiety and fear provoked by the events of the game, all had a greater impression in VR. Because of the device of egocentric picturing, the player is not simply looking at a 2D screen displaying action from which they are distanced, as in traditional horror films, rather they experience the action from a perspective within that world. Things exist and events occur in front of them, to their sides, behind, above, and below: and the player is vulnerable within this space, continually turning their head, listening, terrified of what is in the shadows or behind the door ahead of them.

The plausibility of the special realism of virtual media is also borne out by observations we can make about how people use or respond to the media. While people surely overplay their reactions, some of the naturalness of the responses that VR users have is evident. In a casual, perhaps sadistic, experiment I have conducted on several occasions, I asked someone unfamiliar with VR—usually a friend or relative—to wear a PlayStation 4 VR headset and play through one of the games that is available for the system. I might send them on a deep-sea dive; ask them to drive a fast car; or, if I know they have the robustness of character, suggest they enter a dark and deserted house to see what can be found there. With their headset on, what I typically observe is something like this: seeing a jellyfish float by, the experimental subject reaches out their hand to indicate its physical presence close to them, as if they were trying to touch it; they will sway and lean in the seat as if to counteract the forces of the car they are driving; and if they are unfortunate enough to find themselves in the dark and deserted house, the subject might scream, and rock

DOI: 10.4324/9781003107644-6

back from the disturbing presence that eventually confronts them. It is all relatively safe, because none of these things really exist; but apparently it *seems* real to my subjects on some level.

This apparent realism has not gone unnoticed by academics and scientists. Near the end of his life, with the computer revolution well underway, Ernst Gombrich saw the possibility of virtual reality and its new contribution to pictorial realism. In a 1986 lecture on the nature of the representation of space in Western art, having recently read an article in *Scientific American* on early VR flight simulators, Gombrich concluded, with a certain amount of foresight that, "Maybe one day these technical developments will lead to the rise of a new art form, as did scene-painting in ancient Greece and Brunelleschi's experiment in the early 15th century" (1987).

One of the most familiar scientific investigations of the realism of VR involves virtual pits (Usoh, et al., 1999). In their attempt to provide an objective measure of the subjective experience of "presence," a standard term used to refer to the feeling of realness in VR, Michael Meehan and his colleagues exposed individuals to a virtual "pit room" where they would stand at the edge of a precarious chasm leading to a room twenty feet below. The scientists hypothesized, and found, "that to the degree that a VE [virtual environment] seems real, it will evoke physiological responses similar to those evoked by the corresponding real environment" (2002). The pit experiment now has a popular embodiment in *Richie's Plank Experience* (2017) a VR game where the player is exposed to virtual heights. Furthermore, this apparent realism seems a predicating assumption of many of the therapy and training applications discussed in the previous chapter. The assumption is manifested in these cases by treating VR spiders, heights, open spaces, and so on, as effective proxies for their actual instances. And, finally, there is some evidence that academics tend to think of VR as being a "realistic" medium. The games studies scholar Dooley Murphy has conducted a qualitative analysis of the literature on VR, attempting to substantiate and extend what he takes to be the "common-sense claim that VR 'feels more real' than traditional screen-based games" (2017).

It is thus obvious that VR is often taken to have a *special realism*. Some theorists take the prospects of virtual realism very strongly and see it as an inevitable destination of the project of VR media. The phrase itself—"virtual reality"—is often taken to mean that the medium reaches or aspires to reaching a level of depictive sophistication where what is depicted might be treated *as if it were real*. Inherent in Nick Bostrom's wild claim that we likely already live in a virtual reality is the assumption that intelligent creatures, wherever they exist in the universe, are very likely to develop "ancestor simulations," a kind of virtual reality that employs the technology for visiting past ages and lives (Bostrom, 2003). Under this view, a perfectly realistic VR is an almost universal inevitability. This is a strong assumption indeed.

Let us take a step back and get real about virtual realism. As they are, VR media are very far from the destination of perfect realism. The prominence of their artifactuality will be abundantly clear to anyone who has used them, and impresses itself on the user in a multitude of ways, some of which we have already met here: the screen door effect; limited field of vision; lack of peripheral vision and depth of field; VR sickness; the artificiality of haptic tactile feedback; a weighty sometimes tethered headset; your physical movements constrained to "room-scale"; a crude sense of bodily orientation; limited avenues for interaction; no smell; no taste; no feeling of warmth from the virtual sun on your skin; and which is worst, not being able to see your actual environment and the fear of tripping over the coffee table. Thus, if we are to discuss the realism of VR media here, it is against this background of the medium currently seeming altogether less than realistic. It will be a discussion of how VR differs in realism to other media and is perhaps more realistic in some limited regards than other media.

Moreover, there are reasons to be skeptical about the idea of realism itself, virtual or otherwise. A potential lack of clarity over the nature of realism, and indeed an ongoing disagreement about its nature, is another challenge to the idea of virtual realism. If realism is itself a contested or uncertain concept, this uncertainty could affect the analysis of virtual realism. At worst, we might decide that claims that virtual media are superlatively or uniquely realistic are little more than technological bluster and, given this possibility, that we should be weary of assuming the inevitability of perfect realism. If we are to substantiate the realism of virtual media, then we will need to get clear on just what that realism comprises. As already suggested here, "realism" might mean more than one thing, and these individual senses may be of differing credibility in the case of VR.

6.2 What Is Realism?

What is realism as it pertains to the purported special realism of VR media? The term, of course, plays a role in several different philosophical debates, one of which concerns the ontological status of various items, be they the theoretical posits of science, abstract objects such as sets or numbers, or even everyday objects such as tables and chairs. Such debates have long been a cornerstone of metaphysics, and the debate about the metaphysics of virtual objects clearly does sit comfortably within this traditional concern, particularly when it comes to the views of "virtual realists." This is not the direct concern presently, however.

Another sense of realism, though clearly related to virtual reality, but which again is not the focus of this chapter, is realism in the sense of "practical" or "sensible," in contrast to an optimistic "idealism." Virtual realism is perhaps most closely associated with the pioneering virtual reality theorist Michael Heim. Heim's book *Virtual Realism* (1998) is not principally a work on ontology—though it does include some unsystematic asides on that topic—rather, it is a work on the culture of VR. To be a virtual realist is to have a pragmatic, grounded, and modest orientation on the technology of virtual reality, in comparison to "network idealism" which shares with philosophical rationalism a "faith in [the] progress" of reason, as now embodied in rational virtual technologies (Heim, 1998: 38). "Naïve realists" in contrast have "many fears," including the "fear of abandoning local community values as we move into a cyberspace of global communities," of the "sweeping changes in the workplace and in public life as we have known it," and "of the empty desolation of human absence that comes with increased telepresence" (37–38). Again, this interpretation of realism is certainly relevant here—and I have said almost nothing about how we might value or feel about the "virtual turn" confronting us—but it is not the sense of realism I wish to explain.

Of course, the relevant sense of realism is its use to refer to a work of art as realistic. The concept of realism has long been an important one within artistic history, with some artistic media and styles being plausibly seen as more realistic than others. The Renaissance portraitist Hans Holbein's work seems realistic in a way that the work of many of his contemporaries is not. His portraits have a level of detail and likeness to their subjects that impresses us as being faithful to the appearance of reality. Gombrich notes of Holbein's portraits of the courtiers to Henry VIII, "the longer we look at them the more they seem to reveal the sitter's mind and personality. We do not doubt for a moment that they are in fact faithful records of what Holbein saw…" (Gombrich, 2006: 278). And this is also the sense in which Brunelleschi took his works in linear perspective to be realistic. In Alberti's famous claim that "The painter is concerned solely with representing what can be seen," this realism becomes a matter of the painter's ultimate purpose and relationship to reality (Alberti, 1973: 43).

Hence, I will take my theoretical lead from the treatment of pictorial realism in the philosophy of the arts. It is this sense of pictorial realism, and its apparent desirability, that has underpinned my discussion of the origins of VR as a pictorial remediation of spatial experience. VR media, under the view presented here, have arisen partly guided by the aspiration of an increased, perhaps supreme, realism. But this aspiration has been undertheorized because VR proponents have assumed their medium to be inherently realistic and have vigorously pursued its expansion in a purely technical way, seemingly unaware of the problems and theoretical tensions with the notion of realism itself. One key issue is precisely what this intuitive sense that one picture or style of picturing might be more realistic than another picture or style, amounts to. A first distinction is between the realism of the subject matter and of the techniques used to depict the subject (Hyman, 2006: 191). A unicorn is an unrealistic subject, but even it can be depicted by realistic techniques, and because the intuitive sense of VR realism exists even where the subjects are fantastical, it is the latter sense that is of interest here. So, when is a picture realistic in this technical sense?

A first bad pass at explaining the realism of pictures would be to claim they are realistic when, under suitable viewing circumstances, they are mistaken for the things they are pictures of. In this sense a painting would be realistic when it cultivated the illusion of reality for its viewers. This standard of realism might be appealing to some theorists of VR who see the medium as inherently illusionistic, but this is likely to be an impossibly high standard for most realistic art to meet. There are limited cases such as the *trompe l'oeil* examples discussed earlier which may cultivate a temporary illusion, but the realism that we ascribe to realistic pictures often accompanies the realization of their artifactuality. Holbein's portraits are unlikely to be mistaken for actual people in all but the most unnatural of viewing conditions, but they are realistic, nevertheless. As Hyman notes, "Constable once said that art pleases by reminding, not deceiving" (2006: 207). Similarly, as already suggested here, despite their boosters, while VR media may be perfectly illusionistic in speculative science fiction instances, they are unlikely to meet this standard in any actual cases. I will argue shortly that VR media may not even necessarily aspire to illusionism.

The interpretation of realism as *resemblance* has been particularly prominent in the debate about realism in pictorial media—and pictorial representation more generally—even if it is introduced only to be immediately dismissed. The suggestion is that paintings are realistic when they look like the things they are paintings of. This is a very intuitive notion. However, an immediate and now trite response is the observation that pictures typically do not look much like their subjects: a picture of a cow is a flat surface, whereas a cow itself is rather more cow shaped. Moreover, even a supremely realistic picture of a cow will have predominantly non-cow-like features. We may also wonder whether the analysis is uninformative or even circular: isn't the claim that a painting resembles its subject just a restatement of its realism? The question of realism now becomes, What is it for a painting to resemble its subject to a high degree?

The resemblance account was famously challenged by Nelson Goodman (1976). We encountered Goodman's criticisms of linear perspective earlier, and these are properly understood as a part of his general rejection of resemblance theories of depiction. Regarding realism specifically, Goodman claims that judgments of resemblance change with representational practice so that what is judged as resemblance in one tradition may not be so judged in another. These claims suggest quite a different view of picturing in general, and pictorial realism specifically, in the form of "conventionalism." Conventionalism holds that pictorial realism owes, at least in part, and in some cases the principal part, to the use and recognition of pictorial conventions, and that "the way we see and depict depends upon and varies with experience, practice, interests, and attitudes" (10). There is an aspect of conventionalism in the theory of pictorial realism advanced by Ernst Gombrich, as Goodman notes. For Gombrich, pictorial realism is not some attempt to convey what is seen by the innocent eye, which is itself impossible. It is rather the use of a style to "articulate the world of our experience" in much the way that language is used to impart information (Gombrich, 1960: 90). For Gombrich, this means that picturing depends on stylistic conventions and that there is no single objectively realistic pictorial style.

Goodman's own conventionalism is semiotic in form in that he takes the characteristic symbolic form of picturing to be denotation (1976: 3–6). On this view, like language, all depictive paintings are conventional and what we call realism in paintings are the conventions with which we are familiar. A realistic picture is a picture whose depictive conventions we have fully imbibed. However, this position opens the way for skepticism about realism in a familiar way: if depictive realism requires being conversant with the conventions of a pictorial culture, why suppose that there are any standards of when a picture is realistic? Perhaps realism is entirely relative? Goodman courts just this view when he notes that, "Realism is relative, determined by the system of representation standard for a given culture or person at a given time" (1976: 37). Partly because pictorial modes previously considered realistic can become archaic, the art critic Leo Steinberg is entirely skeptical about the concept of realism and concludes that we can "assert with confidence that 'technical capacity in the imitation of nature' simply does not exist. What does exist is the skill of reproducing handy graphic symbols for natural appearances, of rendering familiar facts by set professional conventions" (1953/1972: 293).

Conventionalism seems like bad news for anyone wanting to characterize VR as a supremely realistic mode of picturing. The potential for conventionalism to collapse into relativity gives us a motivation to reject it, at least if we hold out hope for standards of pictorial realism between cultural epochs, and the special realism of VR. Luckily, for those with these tastes, there are other reasons to reject wholesale conventionalism. Pictorial realism, and pictorial seeing more generally, cannot be entirely conventional because that position could hardly give sense to the phenomenon of new or revolutionary pictorial conventions seeming more realistic than previous ones. The unfamiliarity of VR should make it seem unrealistic, when precisely the opposite seems to be the case. Moreover, some realistic techniques exist independently of artistic conventions in that there is empirical evidence that viewers' visual systems respond quite non-conventionally to pictures, prompted by the visual cues they also find in nature. I covered some of these in Chapter Three when I discussed stereoscopy and other means of allowing for the perception of depth within VR picturing. This again at least suggests that it is something about the visual experiences of pictures and native viewing *matching* in some way that explains the realism of some pictures. As such, VR may again justify Gombrich's observation that "The history of art... may be described as the forging of master keys for opening the mysterious locks of our senses to which only nature herself originally held the key" (Gombrich, 1960: 359).

The idea that pictorial realism can escape conventionalism is given a sophisticated modern defense by John Hyman (2006). Hyman holds that realistic pictorial depiction depends on three features to be found in pictures: accuracy, animation, and modality (194–205). For Hyman, accuracy "combines a degree of individuation or precision with the avoidance of idealization, fantasy, error, and deceit." In this sense, the famous drawing of a flea in Robert Hooke's *Micrographia* is a far more realistic picture than any sketch of a flea I could produce, because it is largely accurate to the detail of an actual flea. Animation concerns the expression of character or emotion in paintings. Hyman notes that an otherwise accurate painting can seem less than realistic because the expressive features of the body of the subject might fail to capture the emotional import of the scene. This is a common failing in the depiction of human faces in videogames, where while the graphical detail and accuracy of the depiction of faces has dramatically improved in recent times, the ability of these faces to express emotion suitable to the situation the character finds themselves in is still rare. Finally, modality is a way of talking about the range of content in a picture, and the kinds of questions it is capable of answering (200). Compare 2D bitmapped sprites to 3D modelled figures in digital animation. In 2D bitmapped sprites of an early generation of computer games, the flat plane of the sprite always faces the virtual perspective of the viewer, and, in some cases, this gives the impression that the objects depicted turn to face the viewer. This is unrealistic, because it is insensible to ask (for other than technical reasons to do with the production of 2D animation) *why* the objects turn toward the viewer. But with the development of 3D animations, it is usually sensible to ask why an object or character turns to the viewer, usually having to do with the motivation of the character depicted. This is because 3D animation is richer in its depictive capacity and subsequently allows a richer perceptual and cognitive engagement: and as such, it is more realistic than earlier modes of animation. In these criteria, Hyman's account of realism is mostly concerned with Western fine art, but in my selection of examples above we can see how it might extend beyond this domain, perhaps to the explanation of realism in VR itself. But note, crucially, that illusionism plays no role here, rather, Hyman presents a plurality of justifications for the realism of one picture or style over another.

Other recent accounts argue that realism may be associated with the quantity, quality, range, or relevance of information to be found in realistic depiction (Abell, 2006; Kulvicki, 2006; Lopes, 1996). For Lopes, pictorial realism has to do with its "pictorial informativeness" in the way realistic pictures capture visible aspects of what is pictured (1995: 277). However, realism does not comprise simply the amount of information in a picture, because cartoons and line drawings, conveying little information, may on occasion seem more realistic than informationally rich pictures (1995: 282). Rather, the realism of a picture derives from its "appropriate informativeness within a context of use":

> In technical drawing for instance, pictures serve to convey information useful for building things, so a system of perspective is used which represents receding edges as "true lengths."
>
> (Lopes, 1995: 283)

Lopes argues that this account "explains the diversity of what is considered realistic in different contexts of use", but also how changes in depictive styles, such as that ushered in by Brunelleschi and Alberti, can seem a kind of "revelatory" realism. In such cases, the context of use undergoes a "radical shift" (1995: 283).

In a similar view, but one that attempts to correct Lopes' view in a way I cannot cover here, Catharine Abell argues that

> pictures are realistic to the extent that they inform their viewers about their object's appearances. A picture that is more informative about its object's appearance will be more realistic than one which is less informative about this aspect of its object. Moreover, because informativeness depends on relevance, pictures will be more realistic the more relevant the information they provide about their object's appearance.
>
> (2006)

Here relevance is defined in terms of the viewer's assumptions about and interest in the world, and thus a realistic picture is one that is a basis for "inferences that could not have otherwise been made" about the appearance of a visual scene or object (2006). Abell argues that this allows for different standards of realism in different communities because of their different interests in objects, but in a way that need not collapse into conventionalism. Thus, on these informational views, perhaps pictures are realistic when they tell us a lot about the appearance of what is pictured, or when in a picture there is an informational richness, such that one can learn much about a pictured scene. Again, these informational conceptions of realism may tell us something about the special realism of VR. One might question whether the "special realism" of VR owes to the "radical shift" of useful context referred to by Lopes, or to its special "relevance" to the assumptions of the users of VR and their "cognitive environment," as detailed by Abell.

Recently, it is increasingly common to see a new diversity or pluralism in the notion of what it is to be "realistic." Dominic Lopes argues that there are perhaps multiple things that we legitimately refer to as the realism of a picture or medium and that there "may be as many conceptions of pictorial realism as there are attributions of it" (2006). There is "lifelike" realism, where pictures "depict things as having features which give them a lifelike appearance," such as "visual detail, weight and solidity, and vitality and motion." In "illusionistic" realism, a picture might be "indistinguishable from the scene it depicts." In "informative" realism, pictures "are realistic to the extent that they are informative about the appearances of scenes." And we might be reminded by "stylistic" realism, that "realism is a property of pictorial styles, as well as individual pictures." This variation might lead us to adopt a pluralism about realism, though Lopes also notes that we might still hope to derive a general account of what the concept of pictorial realism seeks to explain, even though this may result in the understanding that "there may be many explanatory jobs to do, requiring many realisms" (2006). Thus, there is a suggestion that the concept of realism is undertheorized, and is most likely a multivalenced one, but which, if properly analyzed, may reconcile much of the apparent disagreement about the realism of artistic depiction. With this in mind then, let us turn to the issue of realism in VR media.

6.3 Immersion and Realism

In the literature on VR, the special realism of VR has primarily been approached through the introduction of the concepts of *immersion* and *presence*. These are the current ways of analyzing the realism of VR, and this section will argue that these concepts are ambiguous and problematic in a way that makes them unsuitable to this job. Immersion is beyond hope as a measure of realism, though there is more promise in the concept of presence, and so I will need to disentangle the two concepts. Though I have referred to them earlier, these concepts need a basic initial description here.

Most fundamentally, immersion comprises the sense of being taken up completely with an activity and forgetting the world beyond that activity. In Bob Witmer and Michael Singer's standard test of presence, immersion is "a psychological state characterized by perceiving oneself to be enveloped by, included in, and interacting with an environment that provides a continuous stream of stimuli and experiences" (Witmer and Singer, 1998: 227). "Sensory immersion," explains the games theorist Carl Therrien, is "associated with the feeling of being transported to a non-immediate reality in the context of mediated representations. In these cases, it is generally linked causally to the degree of vividness or credibility of the represented reality" (Therrien, 2014: 451). Immersion is often thought to be a central feature of VR and is introduced at the outset as a defining feature, for example in both Heim's and Chalmer's definitions of virtual reality.

The second element usually used to describe or explain the special realism of VR, and the focus of a great proportion of previous investigation of psychological responses to VR, is *spatial presence*. Spatial presence can be characterized as the feeling of being spatially located in the depicted VR space and has been claimed to be a predictable response of standard human perceptual psychology to VR depictions of space and the objects located there. This specific sense of "being there" has also been called "place illusion" by Mel Slater to distinguish it from other meanings that have been attributed to presence (2009). A recent cognitivist account sees presence as rooted in basic and universal human responses to perceived environments—including their aptness for interaction—where "spatial presence is reducible to VR users' unconscious acceptance of an avatar's egocentric reference frame and virtual peripersonal space as viable, supported by the 'tuning out' of contradictory sensory information from the physical environment" (Murphy, 2017: 3). On this view, spatial presence relies on a kind of sensory illusion—or several kinds, because it may not be limited to the visual sensory modality—that are exploited by VR media to give the user the impression of reality.

In the scientific literature, immersion and presence are sometimes credited with an extraordinary realism, as here in some quite striking claims made by Slater and Sanchez-Vives:

> This fundamental aspect of VR to deliver experience that gives rise to illusory sense of place and an illusory sense of reality is what distinguishes it fundamentally from all other types of media. It is true that in response to a fire in a movie scene, the viewers' hearts might start racing, with feelings of fear and discomfort. But, they will not run out of the cinema for fear of the fire. In VR, about 10% did run out when confronted by a virtual fire even though the fire did not look realistic (Spanlang, et al., 2007). In a movie that includes a fight between two strangers in a bar, audience members will not intervene to stop the fight. In VR, they do—under the right circumstances—specifically when the victim shares some social identity with the participant (Slater, et al., 2013), which itself is remarkable because obviously there is no one real there with whom to share social identity. [...] So, VR is a powerful tool for the achievement of authentic experience—even if what is depicted might be wholly imaginary and fantastic.
>
> (Slater and Sanchez-Vives, 2016)

The two examples here are interesting cases of realism indeed, and if true and interpreted correctly, would certainly be surprising to a generation of philosophers who have often claimed that spectators of fictional events do not act in this way, suggesting as they do, the famous thought experiment of Charles and the green slime (Walton, 1978).

In fact, there is no claim in the paper cited by Slater that 10% of people ran from a virtual fire, and this seems to owe to his confusion about what is stated in the paper (which is rather poorly written and conceived). In the second case, it is simply not clear from the described "interventions," which were "verbal utterances or physical moves towards the two virtual characters," whether these were literal attempts to intervene in the virtual fight, or something like a sanctioned make-believe response to the fictional scenario playing out before the participants. It should also be noted that virtual fictionalism has no place in Slater's theory; instead, VR is credited with a "plausibility illusion" and so his participants are responding because the scenario is plausibly real, which is a very strong illusionist interpretation of realism.

To take another example of the potential application of VR realism, in a paper on the potential of "immersive journalism"—a species of virtual documentary discussed earlier—it is claimed that,

> The fundamental idea of immersive journalism is to allow the participant, typically represented as a digital avatar, to actually enter a virtually recreated scenario representing the news story. The sense of presence obtained through an immersive system [...] affords the participant unprecedented

access to the sights and sounds, and possibly feelings and emotions, that accompany the news.

(de la Peña, et al., 2010: 291)

Just how these media comprise an especially realistic mode of conveying journalistic ideas and experiences is not fully explained in the paper, but the justifications such as they are, include a reference to one of the features referred to above in the analysis of realism, namely *fidelity* or *faithfulness* to reality:

> We thus claim that immersive journalism, by allowing for more immersive experiences […] constitutes a much more faithful duplication of real events.
>
> (de la Peña, et al., 2010: 299)

I have seen numerous assertions of the special realism attributable to immersion and presence, in academic work, in the media, and in person: I opened this book with my own frightening experience of the apparent realism of spatial presence in the game *Resident Evil VII*.

Unfortunately, there is ample evidence, and indeed acceptance in some parts of the VR literature, that the concepts of immersion and presence are frequently conflated, and that immersion itself is a deeply ambiguous concept, often encroaching into the territory properly assigned to presence. This makes their conceptual role in explaining the special realism of VR quite problematic. Principally, immersion is vague, falling between being an *attentional concept*— where one becomes oblivious to events and things beyond the activity one is consumed with—and a concept concerning perceptual or spatial experience— where one is "transported to" or "immersed in" another world or is overtaken by a sense of reality inherent in the visual features of a virtual world. The ambiguity between attention and physical displacement is evident in Heim's account of virtual reality and immersion, where he refers both to being "cut off" from surrounding reality and being "submerged" in a virtual world (1993: 112). Witmer and Singer's definition of immersion just cited also courts the ambiguity, by combining the spatial idea of being "enveloped by, included in, and interacting with an environment," and the "continuous stream of stimuli and experiences," which more obviously concerns the flow of attention.

Let us tackle the transportation sense of immersion first. Immersion in this sense is a metaphor, and an unfortunate one. No one is submerged, transported to, or immersed in a virtual world, of course. The *apparent* spatial displacement involved in VR follows from the function of egocentric picturing, which can both place the viewer in another apparent perspective, and allow them, via interaction with such pictures, to interact with things at a distance. It is better to replace these metaphors with the theory of egocentric picturing because it removes this potential source of confusion. As a virtual technology, egocentric picturing no more transports you to another location than does the virtual technology of telephony. Accordingly, this is where presence, disentangled from immersion, plays its proper explanatory role, as presence comprises a psychological response cultivated by egocentric pictures (and perhaps other depictive media).

The attentional sense of immersion itself has competing interpretations, including that the attention derives from the *exclusion* of the actual world—potentially its physical exclusion by the donning of a headset—the attentional "tuning out" of the real world, and the experience of "flow." First, VR seems to exclude current realty by perceptually blocking it out or replacing it with another sensory environment. Heim notes that in VR "the HMD [head mounted display] cuts off visual and audio sensations from the surrounding world and replaces them with computer generated sensations" (1993: 112). Here the attention is focused on the VR experience by simply precluding attention to the actual world. Secondly, the admittedly loose notion of "tuning out" has much the same effect, as the user focuses their attention on the content of the screen, and the sense of presence afforded by it, at the cost of their attention to their actual surroundings. Finally, and perhaps explaining why we tune out in some task orientated cases, is Mihaly Csikszentmihalyi's familiar concept of flow (1990), which is sometimes used to explain the focusing effects of digital games (Takatalo, et al., 2010). Here, the immersion may result from the engagement with the tasks and "challenges" depicted in the virtual world (Ermi and Mäyrä, 2005: 92). Perhaps we need not be committed to flow specifically, as other features of attention allow us to understand how we might become fixated on a task to the detriment of our awareness of the actual world around us. Nevertheless, however it is explained, this sense of immersion refers to the selective attention involved in VR.

As described, immersion is not suitable for capturing the sense of virtual realism introduced above, and that something is immersive does not establish that it is also realistic. As primarily a psychological state—a *response* to VR—immersion may not allow us to say much about the realism of the medium itself. First, many things that are not realistic in the pictorial sense identified above reliably induce attention by excluding the external world or encouraging flow. Indeed, attentional immersion is a factor shared by a wide variety of activities beyond VR, given that Witmer and Singer's analysis applies as much to immersion in the actual world as it does to apparent virtual worlds. But even in the domain of VR, immersion does not seem to be a strong guide to whether an application is realistic or not. Take two VR games that have already been discussed here, *Half-Life: Alyx* and *Tumble VR*: while the latter (as we will shortly find) gives a somewhat realistic depiction of the mass and textures of its puzzle pieces, and the former seems to reach for a pictorial realism not seen in the stylised graphics of *Tumble VR*, these facts have very little to do with the immersive effects of either. In fact, the attentional sense of immersion—the sense of intense focus associated with Csikszentmihalyi's concept of flow—is possible even with non-realistic VR games.

Second, realistic works, even realistic works of VR, need not be immersive in this sense. First, realism can *disrupt* immersion where the events of virtual worlds leave the user feeling especially vulnerable, effectively deterring their participation in the world. One of the most unfortunate side effects of VR—motion sickness—means that rather than being immersive, one must often put in significant effort to tolerate the medium, even as it seems "real"—and perhaps *because* it seems real, at least to the visual and vestibular systems (Lawson, 2014). Immersion, principally, concerns attention or concentration, and while VR media do have attentional effects, this is likely a consequence of the applications of VR media, such as games, rather than a characteristic feature of its realism. Thus, one might even go so far as to say that contra the definitions of VR offered by Heim and Chalmers, immersion is not a necessary component of VR. In some artistic cases, where attention is called to the depictive surface, the medium might resist immersion even as it provides a realistic rendering of a VR subject. Realism may sometimes drag us out of inattentive immersion. Immersion as a term is rather unhelpful in this context, and because the concept is ambiguous, "opaque and contradictory" (Grau, 2003: 13), we may be conceptually better off without it.

6.4 Perceptual Realism

In much of the theoretical and technical literature concerning VR, there is an assumption that the medium is *perceptually realistic*. Perceptual realism here is used to refer to those cases where virtual worlds *seem* to the user as if they were real, or cause their appreciators to respond or behave as if they were real, or engage our perception in much the same way as the natural world does. Moreover, it is the claim that VR media achieve these things because they successfully remediate the experience of space. This is quite a broad brief for the term, but I think we will find that it picks out a category of interest. A notorious case of the perceptual realism of VR is the "virtual pit" experiment mentioned earlier, where participants, asked to approach a pit in a virtual reality environment, are observed to have raised heart rates and other physiological responses indicative of anxiety (Slater, et al., 1995; Meehan, et al., 2002). This assumption of perceptual realism is so strong in some of the literature, that it is often claimed that virtually depicted items or events can go proxy for the real thing in experimental settings (Cheah, et al., 2020; Slater and Sanchez-Vives, 2016). Perceptual realism may also be behind some of the claimed or speculated positive real-world applications of VR, such as its potential in encouraging empathy (Maister, et al., 2015), its use as a distraction from the pain of medical procedures (Indovina, et al., 2018), and the range of therapeutical uses identified in the previous chapter.

Which precise aspects of VR do resemble the natural experience of environments? The *simulative* aspects of CGI mentioned in an earlier chapter—of movement, change, and development, along principled lines—may be an important part of their potential for physical and spatial realism. In particular, simulated physics techniques may strengthen the sense of spatial experience in VR. In *Tumble VR* many of the puzzles depend on the physics of the puzzle objects and environments, and the player's interaction with them. For example, in one series of puzzles, the player must balance objects of different sizes, shapes, weight and constitutive matter so that they reach a given height. Performing this successfully is a matter of arranging the pieces, carefully placing them, but also intuiting the inherent physics of the arrangement. The spatial impression here likely depends on our visual sense of physics, something investigated in recent work employing virtual environments (Cano Porras, et al., 2020).

VR is a form of moving picture, but, unlike cinema, its environments are egocentric and interactive, and so the movement in the picture may be "around" the viewer. This potential of VR to depict a flow of movement around the user may correspond to certain aspects of native perception. One specific kind of perceptual flow are "optic flow fields," in which the animals perceive apparent size and speed of objects in their visual field as evidence of those objects moving closer or further away due to the animal's own movement. Such perceptual flows are "generated by the visual perception of global motion of objects in the environment during locomotion, [and which] are based on the movements of projected images across the retina" (Heesy, 2009: 23). This perceptual phenomenon is crucial to spatial experience, and most animals in which the perceptual phenomenon has been studied, including humans, use it "to determine heading, velocity, and time-to-impact during locomotion" (23). This form of perceptual flow is also in evidence in VR, where it may be crucial to the tasks found in that context: for example, recognizing and shooting at Tie Fighters in the space combat VR videogame, *Star Wars: Squadrons* (2020).

However, virtually realizing this kind of perceptual flow leads to severe challenges for the technology. I noted very early on in this book that one of the chief technological challenges in VR was to produce a continually updated animated environment where latency—the gap between the user's action and its depiction on the screen—was sufficiently low, as high latency disrupts the experience of presence (Welch, et al., 1996). This is sometimes called "motion to photon" latency, and its existence is a contributor to VR sickness. Modern headsets are now able to depict their updating worlds with a tolerable level of latency such that the experience of the egocentric picture may give the effective appearance of the flow of perceived space. This must be a key element of realism, in that the actions of the user, perhaps an action as simple as turning one's head to look in a different direction—essentially, the "sensorimotor contingencies" mentioned earlier (Slater, 2009: 3550)—feed nearly instantaneously into the depiction of the perspectival change, giving a strong impression of visual agency.

The flow of VR experience may be predicated on other aspects of perception, including perceptual constancy and the sense of object permeance, and these too, may count as a way in which VR may be faithful to native perception. Perceptual constancy comprises the ability of animals to perceive the varied and ever-changing stream of sensations they receive as a stable set of properties of objects in their environment. Spatial perceptual constancy has recently been investigated in commercial VR headsets, through the measurement of "the perceptual accuracy of near-field virtual distances using a size and shape constancy task," and the study shows that these VR systems do in fact give their users a realistic and constant impression of the egocentric space they "inhabit" (Hornsey, et al., 2020: 1587). Object permanence is also of interest because this kind of ontological assumption is a precondition of a stable perceptual interaction with any kind of environment. That we so naturally assume that the "objects" encountered in VR are independent, persistent, and tangible things, makes a significant contribution to psychological realism, and, I am tempted to think, to some of the more conscious ontological assumptions that philosophers have made about the apparent objects of VR worlds.

Spatial presence, if disentangled from immersion, might also be thought as a key feature of perceptual realism. However, this is complicated in a way that suggests something important about the perceptual realism of VR. It turns out that spatial presence does not depend on the veracity, fidelity to life, or graphical sophistication of the picturing surface: as Murphy notes, "a user's willing involvement in or absorption in a virtual simulation may compensate for technological [i.e., graphical] shortcomings" (2017: 4). Spatial presence, then, is possible in VR environments that in other respects might not appear to be particularly realistic. This is to be expected, because presence is a quite general feature of VR in both its realistic and stylized cases.

Explaining the existence of presence even in unrealistic VR scenes, is that the realism in such instances derives from a perceptual structure that might be shared by superficially realistic and nonrealistic VR. A frequent argument is that the sensorimotor contingencies associated with viewing VR, that is, the way the tracked perspective of the VR environment allows the user to move their head and body to view the scene, act as spatial cues that prompt the sense of presence (Slater, 2009). Slater argues that it is these kinds of movements that are crucial to the immersion in virtual environments:

> Immersive systems can be characterized by the sensorimotor contingencies (SCs) that they support. SCs refer to the actions that we know to carry out in order to perceive, for example, moving your head and eyes to change gaze direction, or bending down and shifting head and gaze direction in order to see underneath something. [...] For example, turn your head or bend forward and the rendered visual images ideally change the same as they would if you were in an equivalent physical environment. [...] A system that supports SCs that approximate those of physical reality can give rise to the illusion that you are located inside the rendered virtual environment.
>
> (Slater, 2009: 3550–3551)

Hence, the aspect of VR that gives rise to spatial presence is not the faithfulness of the superficial appearance of the depicted environment, but rather the faithfulness of the sensorimotor experience of the perceived space in the ego-centric second fold of the VR depiction. It is the structure of experience and action rather than its superficial appearance that makes a VR system percep-tually realistic in this manner.

This sensorimotor explanation of presence may explain another aspect of the apparent realism of VR. Earlier in this chapter, I noted several cases where users would gesture or move in ways that might suggest that at some level, they "believe" the VR experience to be real. Riders of a virtual rollercoaster will tip forward as the virtual rollercoaster plunges into a descent; viewers of underwater life might reach out to what they see around them. Assuming the actions are not (entirely) pretended, this might seem like powerful evidence of the realism of VR. We now have a possible explanation, combining the sensorimotor explanation of presence and features of the sensorimotor system and spatial cognition more generally. Our physical responses to our environment are not always conscious, with reflexes being largely unconscious. An instance of reflex that will be familiar to most of us is the defensive "blink reflex." Recent research has found that it is possible to generate the reflex with a virtual hand placed in the user's virtual peripersonal space (Fossataro, et al., 2020). Some of the actions I observed in my experimental subjects may result from such unconscious sensorimotor reflexes or contingencies, such as startle or withdrawal reflexes.

But also, the bodily relationship to virtual space may be quite intentional. Just as it is entirely natural to look around and direct one's gaze at the virtual egocentric space, our physical orientation may sometimes lead us to con-sciously gesture or reach out to such environments. In games studies, the interactivity of virtual worlds that I explained earlier in terms of the technical vocabulary of "props" and triggered "events," is frequently explained by referring to J. J. Gibson's ecological theory of perception, and its notion of an "affordance" (Gibson, 1986). Regarding animal perception, an affordance is what the environment offers up as a potential for action: that a surface is "flat," horizontal," sufficiently extended, and "rigid," and "is also knee-high above the ground, it affords sitting on" (1986: 127–128). We might speculate that the connection is not merely terminological, and that Gibson's notion of affordances genuinely explains a user's physical orientation toward the environment of VR, and that this physical connection to the environment may generate something of their apparent realism.

For example, while their account of VR seeing leaves something to be desired in the "ontological shift" they attribute to VR perception, Gra-barczyk and Pokropski do seem to identify something important in our experience of the VR worlds and objects: that they seem most vivid just when and because they embody affordances for action. They note that

> Pictures become objects (environments or worlds respectively) once they start to be perceived as rather persistent, rich (possibly even infinite) sources of affordances...

> (Grabarczyk and Pokropski, 2016: 36)

Of course, pictures never literally become objects, but with suitable modifications, Grabarczyk and Pokropski's observation stands. Traditional pictures, in their non-spatial and non-interactive nature, are quite unlike the more natural aspects of our environment, where seeing is almost continuously conjoined with the possibility of action (Matthen, 2005). The interactivity of VR egocentric pictures may introduce a new realism into picturing by embodying a physical, and indeed, *bodily* relationship to depictive spaces (Slater, 2009: 3554). And the partial illusion fostered by such pictures, may naturally engage the sensorimotor contingencies that lead to behaviors such as swaying, leaning, and gesturing toward objects apparent in the virtual space.

Finally, there may be aspects of our affective responses to VR that fit this analysis of perceptual realism. The emotional response of fear I experienced while playing *Resident Evil VII* was one reason that I found the VR version of that game more impactful, and perhaps, more realistic. Assuming the perception was accurate, what is the difference that allows the VR version of the game to have this greater impact? The answer, I suspect, is that because VR is an egocentric and interactive picturing form, it may better depict the experiential contexts of emotions elicited by the gameplay. This begins with the sensorimotor contingencies noted above: because the player is placed within the environment, they derive information about the environment through shifting their head and eyes to look, bending down, and perhaps even leaning around a corner. When the player is placed within the environment in the context of a survival horror game, this situation brings a sense of vulnerability and urgency to the informational needs of the player, so they might find themselves nervously looking over their shoulder or leaning around doors, afraid of what might lurk in the dark. These perceptual contingencies are both an expression of, and coping response to, the fear of the environment.[1]

In his review on the literature on player experience in VR, Dooley Murphy finds that a predominating theme in the phenomenology of VR is what he calls a sense of "patiency," that is, a feeling of having limited agency in the VR world, and a "co-occurrent [...] sense of self vulnerability" (2017: 10). Hence, emotions of fear and anxiety might be very natural in a VR world as they are frequently elicited by insecurity or environmental threats. Finally, within an interactive environment of VR, these emotions have an available means of behavioral expression they may not have in other media settings. It is unlikely that users of VR would run from a virtual fire; however, it may be possible for them to *fictionally* run from a fire if the VR application allows for this. Fight or flight are concomitant behaviors of fear, and these responses can easily be incorporated in the affordances of a videogame, with that medium's frequent fictional manifestations of aggressive or violent play (Tavinor, 2017). In my earlier book on videogames, I argued that the interactive worlds of videogames might modify the kinds of emotional responses appropriate or possible, allowing users to have self-directed emotions (Tavinor, 2009). We can now see that the same is the case of emotions in the context of egocentric pictures, though with the added richness of experiencing the fiction from within the graphical environment. One source of the affective realism of VR, then, follows from the behavioral expression of emotions, such as fear, being possible in an egocentric interactive picture.

We can now also see that it is the use of VR to remediate spatial experience, and the potential fulfillment of this intention, that can be interpreted as its chief claim to special realism. Earlier I suggested that the design of VR might frequently seem like a return to the realist aspirations of artists of earlier times. In fact, VR seems to boldly confront many of Goodman's complaints about resemblance and the realism of linear perspective. The utilization of VR headsets and motion tracking makes it false for VR depictions of spaces that "the picture must be viewed through a peephole, face on, from a certain distance, with one eye closed and the other motionless," because VR allows for binocularity, the scanning of visual scenes, and the apparent movement of the viewer into the visual scene, and their bodily responses to the spatial experience, all in a way that *resembles* the flow of natural perception and action. Yearly there is an increasing array of such representational techniques, and what the future will bring may truly astound us.

We have found that resemblance is a repeated theme in the explanation of realism, and we might now conclude that this perceptual realism can at least be interpreted in terms of resemblance and fidelity. More strongly, it might be thought that, if it could successfully respond to the kind of conventionalist arguments framed against resemblance theory, VR might *rehabilitate* resemblance as a criterion of pictorial realism. For many theorists and designers of VR, it is thought to be crucial that VR environments do resemble the ordinary world, and fidelity is frequently conceived as an overt aim in the design of VR depictions. As cited earlier, "it is important for the graphic image to create a faithful impression of the 3D structure of the portrayed object or the scenes the displays depict" (Hoffman, et al., 2008: 01) and ideally, VR must "convey to the observer the illusion of being in a complex structured space of a natural world" (Grau, 2003: 15). This sense of realism as resemblance is also inherent in Therrien's account of sensory immersion when he refers to the "vividness or credibility of the represented reality" as contributing to the sense of immersive transportation (Therrien, 2014: 451). It is also evident in Slater's notion of the "plausibility illusion" (Slater and Sanchez-Vives, 2016).

However, it should be noted that what is referred to as resemblance here is a little different from the candidates for resemblance in earlier debates regarding pictorial realism. Ignoring for the moment the problems with previous resemblance accounts of pictorial realism, if one were to claim that a Holbein portrait resembled its subject, this would be a claim about the faithfulness of the *appearance* of Holbein's painting to the appearance of its subject. Because VR is an egocentric and interactive form of picturing, the resemblance or fidelity in VR amounts to the claim that the structure of perception and action in VR media resembles, or is faithful to, natural perception and action. This resemblance is hence a sensorimotor and perceptual example of a *structural isomorphism* argued in Chapter Two to be amongst the characteristic features of virtuality. The claim, then, is that VR is realistic because of the perceptual isomorphism between the experience of the actual world and that of apparent virtual worlds.

The rehabilitation of resemblance as a measure of pictorial realism would be a significant development in understanding the potential of pictures. VR replaces the static, insensitive, non-interactive spaces of paintings and photographs with the dynamized and encompassing visual field of an egocentric picture. And in doing so, it may genuinely resemble the scenes it depicts. Under this view, the faults that led to the critique of the theory of resemblance were not in the standards of resemblance or fidelity themselves, but owe to a failure in the kinds of pictures to which these standards were applied. Previous modes of picturing, in their spatial paucity and lack of dynamization, are just unrealistic compared to the special realism of VR. This is a very bold claim indeed!

6.5 Is VR Realistic?

If the realism of VR can be ascribed to its remediation of the perceptual structure of our natural experience of the world, and of resembling or being faithful to an experiential and agential situation within a world, we have a way of assessing how realistic VR is, and we have at least some evidence to think that VR media do extend the possible realism of picturing. Nevertheless, we can also be critical about these claims, and this criticism is crucial given the early stage of this debate, the evident hype and boosterism around VR, and that so much of the study of VR seems to assume without question the special realism of the medium. In recent work, Janet Murray thoroughly admonishes some of the more extreme cases of hyperbole around the putative indistinguishability of VR from reality. She notes, quite sardonically, reports of a presentation at the world economic forum in Davos where "white men in white shirts, suits, and ties with [VR] headsets on" found themselves virtually in a refugee camp in Jordan: "the claim that billionaire bankers would experience a fundamental expansion in human sympathy by viewing a 360 film through a headset is pure wishful thinking" (2020: 13).

Here too, I want to make the case that the final claims of the previous section are surely a little too bold. Thus, in this section I am going to sketch several kinds of criticism of the virtual realism described in this chapter. First is that VR provides only a very incomplete approximation of our perceptual engagement with a world. Second are the practical problems with thinking the task of remediating experience could be completed. Third, is the criticism that the idea of spatial remediation might be deeply confused because of its assumptions about perception itself. Fourth, and related, is the critique that the apparent VR realism presented here comprises a set of conventional features, and that there is reason to be skeptical of their naturalism. Finally is the question of whether realism is an inevitable goal of VR, as some technologists and academics seem to have assumed.

First then, let us return to the caveat I made at the beginning of this chapter that any discussion of the realism of virtual media needs to acknowledge that virtual media as they currently are, are not particularly realistic, especially if, as we can now see, that realism is conceived as the remediation of natural experience. This needs to be emphasized. VR works fail to be realistic in many more ways than they achieve realism. And so, the special realism described here is only relative to other media: we may be right to judge that VR media are more spatially realistic in their portrayal of egocentric space *than previous kinds of picturing*; they better present the flow of perception and its continual conjunction with the possibility of action, *than previous modes of picturing*. And so on. But on their own terms, VR media are a very incomplete and poor remediation of our experience of the world. They are still in most cases largely confined to the senses of vision and hearing, and even in their remediation of these sensory channels they are severely compromised. The field of view is unnaturally limited; there have only been very crude attempts to simulate depth of field effects; and the actual rendering and animation of the perceptual content—the environments, objects, and inhabitants one encounters in virtual worlds—is all too evidently artificial. These shortcomings undermine not only the idea that VR could be truly illusionistic, but also the less demanding claim that they are particularly faithful to native perception.

Moreover, virtual media are likely to remain rather less than fully realistic for quite some time. Vision and hearing are the most common sensory channels conveyed in VR experiences, and the reasons for this is perhaps that they are the most tractable: both involve a reasonably simple sensory surface, at which stimuli can be directed to give the impression that the user is receiving information from an actual world. And the VR industry is already working to improve some of the omissions in these sensory channels, in the form of virtual sound, eye-tracking, multi-planer displays, and foveated rendering. The sense of smell, perhaps, could be virtualized rather simply—but one wonders if VR users really would want to smell some virtual worlds! But other sensory channels might not be so simple. Tactile and haptic feedback are a challenge. The current haptic gloves covered in an earlier chapter are extraordinarily crude and give only a very vague impression that one might be touching and holding a physical object, mostly through "force feedback": that is, small servos restricting the movement of the fingers when they encounter a virtual solid object. In these cases, tactile impressions are entirely missing, as one has no feeling of the surface of the object. This is a practical problem, and perhaps we might imagine how a peripheral mechanism could simulate different surfaces for the fingertips, so that one could discriminate between a smooth wood of the guitar neck you are virtually playing, the soft fuzz of a peach you are about to virtually eat, or the sharp scalpel that has accidentally just cut into your flesh. Compound this by extending the sensory area to the surface of the whole body, and the engineering challenge of how to intervene at this sensory surface to give the virtual impression of a tactile environment seems almost inconceivably difficult.

There are, of course, far more than five senses, and the VR literature that I surveyed in the preparation of this book sometimes seems oblivious to these sensory variations and the engineering challenge they would present in the search for *perfect* VR realism. The extended list of sensations, including pain; hunger; the vestibular system; the sensation of the dilation of blood vessels that may be experienced as a headache; sensory receptors in the esophagus that give the feeling of swallowing (and vomiting); the cardioception involved in the monitoring of the heart; and sensory receptors in the gastrointestinal tract that monitor distension and indicate flatulence and the urge to defecate, would all need to be represented in perfect VR, and all would present engineering challenges in the production of a depictive "surface" that would allow for the virtual experience of these sensations. And at least some of them give me a small shudder of trepidation!

The obvious way to avoid all these practical problems, of course, would be to circumvent the sensory surface altogether, and place the point of technological intervention within the nervous system itself. What we need for perfect VR is not an intervention at the sensory surface, but a brain implant that feeds simulated transduced sensory information directly into the central nervous system. This, as I acknowledged earlier, has been the topic of recent speculation amongst futurists, technological entrepreneurs, and related species of fabulist (Regalado, 2020). Nevertheless, this might seem like an elegant and simple solution to the engineering problems of intervening at the sensory surface, but it only does so because we—and I am referring here to the reader, and the VR technologist—currently have little comprehension of the feasibility of this project: the apparent elegance derives from our current ignorance of the engineering challenges involved, and perhaps, our current technological hubris. This project, as it stands, is science fiction. But even if science fiction, it may be merely a problem of practicality, as previous science fiction scenarios have recently had a habit of being realized by technology. Other non-practical problems loom, however.

This science fiction speculation about the future of the VR medium involves some assumptions that should lead us to a third, conceptual, problem facing the VR remediation of perception. The idea of a brain implant feeding transduced representational information directly to the central nervous system may involve certain computational or representational assumptions about the workings of the brain, that is, that it does work on principles generally commensurate with the computers that we have designed over the last century. These assumptions are not philosophically uncontroversial because computational theories of mind (e.g., Putnam, 1967) and representational theories of mind (e.g., Fodor, 1975) have their detractors (e.g., Putnam, 1988). The whole debate about the substrate of the mind gives the impression of being an immature one, even years onward. Fortunately, whether the brain is computational or representational in nature is not my interest here, rather, it is an example of the assumptions that underlie the technological ambitions of some in the world today.

The problems here are a little hard to illustrate and intuit, because in part they depend on assumptions we may not know we are making—a species of former American Secretary of Defense Donald Rumsfeld's notorious "unknown unknowns"—so let us return to the historical case of Brunelleschi's attempted remediation of spatial perception in the form of linear perspective. To capture the geometry of the Florentine Baptistry, Brunelleschi had to make an *idealization* of native perception: the view was to be seen through one stationary eye. But natural vision is ordinarily stereoscopic, and so much—though as we found, not all—spatial perception involves not one but two retinal images, which register the object from slightly different vantagepoints and as having different (though perhaps subtly different) spatial configurations. In perception, these fuse into a single coherent object, but there is often little evidence in normal perception, or at least it is not obvious to us as perceivers, that the single object we see is a combination of different apparent geometrical objects. Evidence of this can be noted by fixating with one eye on how an object occludes background features, and then by switching eyes. Nevertheless, the combination of these two images in normal perception into a single coherent perceptual object already suggests that native seeing is not a matter of perceiving the exact geometry of the light waves that travel from objects to our eyes: rather, the perception of objects sometimes obfuscates precise geometry to suit our informational needs (that is, interacting with singular unified objects such as apples and cigarette lighters). As such, Brunelleschi's attempt to remediate what is seen (if this was his intention) was based upon the faulty assumption that perception does provide us with a simple picture of reality. The question now is, do other such perceptual obfuscations, perhaps related to the flow of spatial perception, suffuse the "simple" perception of the environment that VR assumes in its attempts at remediation?

This leads us to the fourth problem with virtual realism, and it is one that draws directly on the historical precedent of the debate about depictive realism. Earlier here, virtuality was characterized as sitting between the poles of "replication" and "representation." A virtual item like a piece of currency cannot remediate the item in a form too similar to the original, because it would merely replicate it: coinage produced in a different alloy would remain actual and not-virtual coinage. But it must remediate the function of the original, otherwise it threatens to become a mere representation of currency, such as play money. But, in many of the cases discussed above, it seems evident that VR media tend toward the latter pole by merely providing token representations of elements of experience, and hence not remediating them at all. This is perhaps most obvious in the case of haptic feedback: the buzzing of a controller does not so much remediate the experience of feeling an explosion nearby, but merely *indicates* or *denotes* that such an event has occurred in the game (given the conventions surrounding the previous use of haptic feedback). In haptic gloves and similar devices, the experience of the haptic surface does not particularly resemble the remediated experience but stands as a representative token of it: one needs to understand that, by convention, a buzzing feeling is representative of a tactile encounter.

Hence it remains a possibility that VR media and their use to depict objects and worlds are not a remediation of perceptual experience at all, but a conventional set of signs that are taken to denote the experiences, objects, and worlds they putatively "virtualize." What evidence, beyond the interpretation of haptic displays, might we muster in support of this claim? Are there other ways that VR purports to realistically remediate a perceptual experience but on examination may merely represent by denotation? Stereopsis gives a potential example. The visual impression of spatiality in VR is often credited as relying heavily on stereopsis, and the feature was central to my earlier account of egocentric picturing, at least as this involves stereoscopic headsets. But in natural contexts stereopsis is responsible for only a portion of our native visual perception of depth, and it operates best at relatively short scales (Heesy, 2009: 30). Stereoscopic headsets may give an especially strong spatial impression just because their images are designed to produce retinal disparity in a way that aids stereopsis. In natural situations other visual evidence such as motion parallax and the relative sizing of objects contributes much to depth perception. If this is true, then the impression of spatiality in VR may depend on the manipulation of stereopsis to produce a striking but ultimately unrealistic effect. And, on inspection, there often is something unnatural about the impression of three-dimensionality of VR in that it seems *too richly spatial*. Spatial presence, the aspect of the experience of VR that is most frequently identified as responsible for its realism strikes my eyes at least to be quite unrealistic, almost symbolic of the impression of actual spatial experience, but not corresponding to the genuine structure or appearance of native perception.

This interpretation of stereopsis as a symbol and not remediation of native perception fits with some applications of VR. I argued earlier that some VR spaces were simply impossible for an actual spectator to view. Examples are of two kinds: first, where VR presents impossible places, such as in the game *Superliminal* where the depicted spaces are like virtual M.C. Escher paintings, and secondly where VR presents impossible acts of viewing, such as traveling though the universe at well above the speed of light, or viewing things at a subatomic scale. That VR can represent perceptual experiences that are impossible for actual viewers—that is, *visual paradoxes*—may imply that VR is much more like a conventional language where words can be combined to describe and seemingly denote situations that could not possibly exist. It may be that VR similarly comprises a merely conventional set of signs that are combined in such a way that we *interpret* the visual reference of the pictorial elements, rather than see through the surface to the elements depicted.

I am clearly reaching here to offer an alternative and critical view of VR realism, but it is at least conceivable that what many take as the realism in VR is instead a set of conventions that have been developing around the use and potential of egocentric pictures. That VR is visually striking need not be in doubt, but this does not mean that we also must accept that it has a *special realism*. The feeling of presence may be more like a striking perceptual illusion that we can cultivate through the manipulation of stereopsis by pictorial elements discovered to have this effect, rather than anything remotely realistic in the sense of resembling the experience of reality. In this case, VR would not resemble our experience of reality but instead offer a set of conventional techniques for denoting that experience, but which may also strike us as sensorily rich and engaging in their own distinctive way. We might arrive back at Steinberg's skeptical claim that depictive realism amounts to "the skill of reproducing handy graphic symbols for natural appearances, of rendering familiar facts by set professional conventions" (Steinberg, 1953/1972: 198). Moreover, it is undeniable that there *are* conventions in VR, as in the case of the haptic techniques just mentioned. Perhaps the medium is thoroughly conventional in the way that Goodman and Steinberg claimed of traditional picturing?

The debates about depictive realism did not end with Goodman's conventionalism, and in a previous section we encountered informational and pluralistic conceptions of realism. Even if the resemblance of the experience of VR to native perception cannot be assumed, and even if it is somewhat a conventional mode of depiction, it may be that other measures of the realism of VR are possible. Because of their very nature, egocentric pictures may align nicely with the informational conception of realism: egocentric picturing is a function of the configuration of the pictorial surface providing the viewer with spatially counterfactual information about the objects depicted, and this information can be employed by the user to perform tasks not possible in other picturing modes. Similarly, the proprioceptive feedback from a haptic glove can provide additional spatial information to its users, and this is why such things are useful in providing feedback to surgeons in their training (Van der Meijden and Schijven, 2009). The physics of depictive animations may give us more information about the physical behavior of the objects depicted, and this may allow the user to catch a ball (Pan and Niemeyer, 2017). But this need not mean that such experiences truly resemble the experiences they denote. Pluralistically too, we might consider VR to provide increments of realism over previous depictive forms. There is "lifelike" realism to some cases of VR, perhaps comprising the "visual detail, weight and solidity, and vitality and motion" (Lopes, 2006) that VR can depict of the objects that one finds in their egocentric space. And there may be at least an aspect of "illusionistic" realism in VR, where the medium, briefly perhaps, gives a sudden impression of real life.

Finally, there is the question of whether realism need be the goal of VR. When Grabarczyk and Pokropski ask, "Is the whole point of VR not that the experiences it produces are indistinguishable from the experience of real things?" (2016: 27) the question demands to be asked in a non-rhetorical way (as most such questions do). Is or should it be the aim of virtual reality to produce the experience of illusionistic or realistic worlds? This kind of ultimate realism may be one aim of technologists, but it is surely not the entire potential of the technology, and it does not exclude other uses to which VR media might be put. And many cases of VR do not aim for anything like realism. *Ghost Giant* as a game has few aspirations to remediate spatial experience in a realistic manner, instead seeking a warped and cartoon-like style. Other applications of VR's egocentric picturing distort spatial experience in surreal ways: *The Cubical* (2016) is a short and disturbing VR rumination on the experience of office work. *Virtual Virtual Reality* (2017) plays with the concept of virtual reality itself to allow the user to flit between nested and looping virtual worlds, often in a way that is spatially disconcerting and unnerving. And other VR applications do not even try to remediate any kind of perceptual experience but employ the technique of egocentric picturing to aesthetic effect, or for the point of spatial abstraction. Many of the works produced via Tilt Brush and similar apps display this intention. The wonderful works of the VR artist Scott Bennett (aka scobot) present richly colorful abstract VR spaces that are an extraordinary pleasure to view.[2]

In these cases, the idea of depictive realism seems rejected in favor of what can be achieved stylistically and artistically with the medium. This too, is a common impulse in art. Immediately following the period of the High Renaissance in Italy was a period in artistic history that we now know as *mannerism*. In mannerism, much of the realism inherent in the earlier work seems to be rejected: the portrayal of human anatomy, for example, became very unrealistic in that human bodies are distended and posed in ways that often make little physical sense. Four hundred years later, at the turn of the 20th Century, depictive realism was again under fire when the analytic cubists fractured depictive space into a set of flat planes. In a work like *Gefect* (1930) Paul Klee reduced the pictorial space to vectors in which the objects depicted—battling people—decompose into primitive edges and color fields. So, drawing our attention back to VR, why should we even think that it will always be used with realistic intentions in its future use?

Perhaps even the "serious" or practical uses of VR do not require substantial realism. In virtual transparency, though the ball being caught might be made to look real via the use of sophisticated graphical techniques, it need not. The transparency of VR media does not necessitate depictive realism because even distorted images or images lacking detail can allow us to see the objects they depict. And this is what we find in the Disney experiment as it was actually conducted: the ball, environment, and the user's own body representation are minimally detailed and quite unrealistic (Pan and Niemeyer, 2017).

The conclusions made in this chapter, such as they are, are necessarily tentative. It is too early to conclude on whether VR media remediate native perception, whether they comprise a set of conventional signs, or whether realism should even be an overriding concern of VR. The arguments or positions scoped above may not be convincing, but these are early days in the aesthetics of VR and there are clearly promising issues to follow up here. The advances of VR depiction certainly have the potential to revivify the debate about the realism of depictive media. Moreover, this debate is worth entering into because the scientific and academic discussion of VR has so often been oblivious to the possibility that VR media are not supremely realistic or illusionistic. Far from considering VR as a picturing medium, with the inherent questions about the realism of that medium, some writers jump immediately to the conclusion that VR comprises a kind of "alternate reality" or more simply, is "real." It is finally time to turn my attention to those accounts of VR that see it, not as a potentially distinctive kind of picturing medium, but as a kind of reality.

Notes

1 Albeit, they are responses to a purely imaginary environment, even if it is embodied in the richly perceptual and interactive manner of VR. Many readers will be aware of the considerable debate about the status of the emotions experienced in conjunction with fictional works (for example, see Walton, 1990: 241–255). There is not space here for me to enter that debate or make any specific observations about how VR fictions may sit within it.
2 A selection of scobot's work can be found at https://www.scobot.com/blog/

7 Virtual Reality and Ontology

7.1 The Temptations of Metaphysics

It has been my contention here that when philosophers have approached the issue of *virtual reality*, they have often, if not principally, focused on the latter part of the coinage to orientate their studies toward metaphysical issues. This is an understandable place for philosophers to begin with the topic, as philosophy has had a concern with virtual realities long before they had that name. Descartes' malicious demon, Berkeley's idealist world view, Robert Nozick's experience machine, and Gilbert Harman's brain in a vat all conceive of possible worlds (or in Berkeley's case, our actual world) as comprising not the real, physical, and material place we naïvely take it to be, but as some kind of illusion or ideation. It is entirely natural to see VR as a technological consummation of these thought experiments, and VR is an incredibly fertile ground for raising such issues. I have found in my own introductory philosophy courses that virtual worlds of the kind the students may encounter in videogames and popular media are an ideal means of introducing the ideas of skepticism, mental representation, ontology, realism and so on. The stunning technological leaps of the last twenty years only make this an increasingly natural approach to the topics.

As I have noted here, these philosophical accounts of VR often begin, not with the incomplete and rudimentary virtual media that we now have access to and that have been the subject of this book, but with the hypothetical idea of a "perfect virtual reality." A perfect virtual reality is a simulation that is sensorially replete, in providing its subjects with all the modes of sensory experience lacking from current VR media. It is also a virtual world that cannot be identified as such from within. Of course, the greatest philosophical precedent for the idea of a perfect virtual world comes from René Descartes, when he closes the first of his *Meditations* by considering that

> some malicious demon of the utmost power and cunning has employed all his energies in order to deceive me. I shall think that the sky, the air, the earth, colours, shapes, sounds and all external things are merely the delusions of dreams which he has devised to ensnare my judgement.
>
> (1641/1984: 15)

DOI: 10.4324/9781003107644-7

This demon threatens to deliver Descartes into a perfect virtual reality where the sensations, and the world perceived through these, do not correspond to an external reality. This thought, propelled down the generations by any number of introductory philosophy courses, has resulted in various pop cultural manifestations such as the "holodeck" of *Star Trek: The Next Generation*, and the movie *The Matrix*. In the former, the role of the demon is taken by benevolent Federation scientists, aiming to provide the crew of the Starship Enterprise with entertaining diversions; in the latter, intelligent machines of some form use the perfect virtual reality to enslave humankind, though this gets a little bit confusing as the film series continues. It should be noted that the Matrix is a slighty less than perfect world because glitches can occasionally be perceived that betray its simulative nature.

Nozick's famous "experience machine" thought experiment gives a classic formulation of a perfect virtual world:

> Suppose there were an experience machine that would give you any experience that you desired. Superduper neuropsychologists could stimulate your brain so that you would think and feel you were writing a great novel, or making a friend, or reading an interesting book. All the time you would be floating in a tank, with electrodes attached to your brain... Of course, while in the tank you won't know that you're there; you'll think it's all actually happening.
>
> (Nozick, 1974: 42)

Nozick employs the idea of a perfect virtual reality to rouse his readers' intuitions over whether there is more to human good than subjective happiness, but the thought is fascinating on its own terms, and has spurred a great deal of thought and argument.[1] Perfect virtual worlds are thus common both to pop culture and academic thought, and embody a persistent philosophical intuition that probably predates even Descartes.

Set against this background, it becomes natural to consider VR in ontological terms, and to take the key questions regarding VR to involve the ontological status of virtual worlds and objects. Are VR objects real? Or are they merely imaginary? Who are the subjects of these virtual experiences? What is the value properly associated with such virtual objects and experiences? Some of these questions, no doubt, will have informative answers, though others, I suspect, involve false dilemmas; but all instantly orientate the discussion toward explaining the ontology of VR objects and worlds.

Moreover, this orientation implies a set of positions on VR, depending on how one answers these questions: *virtual fictionalism* makes the claim that VR is a kind of fiction, and that its objects and worlds do not really exist. Examples include the games studies scholar Jesper Juul's claim that "VR has fictional aspects all the way down" (Juul, 2019: 4) and Neil McDonnell and Nathan Wildman's conclusion that VR can profitably be understood as a form of Waltonian fiction (2019). *Virtual realism* is the claim that apparent virtual objects are real in some sense, though what this means is defined in different ways by different realist theories. In Chalmers' theory of "digitalism," "virtual objects are real [digital] objects, and what goes on in virtual reality is truly real" (2017: 309). Phillip Brey claims that "virtual apples simulate or imitate real apples" (2014: 43). Peter Ludlow argues "that virtual 'physical' objects like tables and plates of sushi just are social objects" (2019: 2–3). And Klevjer's quite elaborate position has it that "*virtual* objects are, at one, algorithmic entities and tangible objects. This dual ontology carries associations to the scientific concept of nature as information, nature as code" (Klevjer, 2017: 733). I am going to refer to the various positions that see virtual objects and worlds as somehow "real" and not "imaginary" as instances of "ontological realism," to distinguish this issue from the virtual aesthetic realism discussed in the previous chapter.

Unfortunately, the metaphysical orientation on virtual worlds and objects, though it is beginning to bear fruit in the form of an active literature, has not been altogether positive. First, this tendency to take a metaphysical stance on virtual reality has to a great extent excluded the consideration of the kinds of questions I have addressed in this book.[2] This ontological orientation is a trap, and most philosophers now working on VR are thus willfully ensnared in a way that has been to the detriment of our understanding of actual VR. Secondly, I believe that the orientation has simply led to a great deal of confusion about the topic, and what we will find in this chapter to be some very surprising and implausible claims about the metaphysical import of VR, perhaps most notoriously in Bostrom's argument that our actual world is very likely to be a perfect virtual world (Bostrom, 2003).

My intention throughout this book has been to stall the consideration of metaphysical interpretations of VR until my full theory of VR was in place. This was firstly, so as not to become sidetracked by the above questions and concerns, and second, because I believed that once the full theory of VR as a medium was in view, the motivations for metaphysical responses to VR would be significantly undermined. In this chapter, then, I am going to spend much less time, effort, and detail on the task of arguing against the metaphysical approach than some might expect. This is because I take the theory presented in this book to have already undermined the metaphysics of virtual reality, and that what is needed here is merely a joining of the argumentative dots. Hence, the focus and approach taken throughout this book—on how people use virtual media to do things, from interacting with the real world, playing imaginative games, making art, and a whole host of activities besides—also has the effect of showing that metaphysical accounts of VR are redundant.

When a philosopher claims to have a distaste for metaphysics, one can typically find them making metaphysical claims. And so, my own position has its metaphysical outlook, and I have here presented a metaphysics of VR, albeit one that does not see the need for too much in the way of additional ontological commitments beyond those that many of us already sign up to. In Chapter Two I spent a fair bit of time discussing the ideas of media and virtuality in a way that does commit me to certain ontological views, particularly about functions and kinds. I do not think that anything in the book is all that extravagant, though some of it, of course, depends on speculative *factual* claims about perceptual psychology. This was part of the rationale for presenting VR in an historically contextualized way: by showing VR to be connected to, and growing out of, precedent media forms such as picturing, the idea that we need a special theory for VR, with special ontological commitments, has been undermined.

The claim that there is a lot of metaphysical confusion about VR might seem unnecessarily provocative. Philosophers argue in good faith about VR, and we should be charitable to their arguments. So, in that spirit, what kinds of arguments can be given for the claim that VR has an ontological significance? In this section I will cover several metaphysical interpretations of VR. Though these arguments are far from exhaustive of all the positions that have been taken on the topic, they are amongst the most well justified and compelling of such claims. If my theory is informative about the failings of these accounts, I think we can assume it will also cope with many of the less well-argued forms of metaphysical realism about VR.

7.2 The Mixed Reality Spectrum and Ontological Shift

Obviously, I need to show that ontologically realist accounts of VR do indeed exist and that I have not presented a strawman. Some of the cases do not explicitly sign up to ontological realism but make claims that can hardly not be interpreted in that way. Other positions, though, are fully onboard with the metaphysical outlook. Let us begin with some of the less explicit cases of ontological realism concerning VR. I noted earlier that there is a frequent and unfortunate notion that reality, augmented reality, mixed reality, and virtual reality belong on an ontological continuum arrayed from more to less real. A Microsoft webpage titled "What is Mixed Reality?" published in 2020, contains a diagram where two labels are attached on either extreme of a graded line. On the left is the "physical world," and on the right, the "digital world," and bridging them are the labels "augmented reality" and "virtual reality." The whole arrangement is labeled the "Mixed reality spectrum."

This characterization frames virtual reality and augmented reality as grades of reality standing between the physical and the digital. While of course this account is not a fully academic or philosophical one, it does capture something of a prevalent view of the metaphysics of VR. At either end of the ontological spectrum is a grade of reality: on one end, the familiar and comfortable real world, and at the other the exotic digital world, and between these, mixed, augmented, and virtual reality with the implication that these differ largely with respect to the nature of their intentional objects, that is, what their representations ultimately denote. This conceptualization rests on the work of human engineering scientists Paul Milgram and Fumio Kishino where it is presented under the moniker the "virtuality continuum" (1994). For Milgram and Kishino,

The concept of a "virtuality continuum" relates to the mixture of classes of objects presented in any particular display situation [...] real environments, are shown at one end of the continuum, and virtual environments, at the opposite extremum. The former case, at the left, defines environments consisting solely of real objects [...] and includes for example what is observed via a conventional video display of a real-world scene. An additional example includes direct viewing of the same real scene, but not via any particular electronic display system. The latter case, at the right, defines environments consisting solely of virtual objects [...], an example of which would be a conventional computer graphic simulation.

(1994: 3)

The clear implication in all of this, is that there are different kinds of object—real and digital—and characteristic means of displaying these: in the former case, conventional video displays or even native perception, and in the latter, augmented and virtual reality techniques.

It is problematic to think of virtual reality and reality arrayed along such a continuum because this formal arrangement conflates the *medium of a representation*, and the *intentional object of such a representation*. I have argued that virtuality is a means of remediating experience, both in terms of perception and interaction, and it most frequently does so through egocentric interactive pictures. But such media have a broad range of uses, and so a variety of intentional objects. While it is frequently the case that the apparent intentional objects of VR are fictional, we have found that in documentary and transparent uses of VR media, they may have as their intentional object real items, even items that are physically present to the user. It is hence wrong to bundle up the medium of VR with its characteristic objects in the way that the mixed reality spectrum does. The two aspects need to be disentangled, with virtuality properly seen as a fact about VR media and their modes of depicting objects, and the apparent ontological issues here amounting to a variation in the intentional objects of such pictures and other representations. This conflation of the medium of VR and its intentional objects is perhaps the persistent confusion in ontological accounts of VR, as we will see below.

The mixed reality spectrum illustrates another of the problematic aspects frequently found in ontological realism, this time, involving claims about the role of *belief* in the experience of VR. This is evident where the author of the document also provides a gloss of the nature of the continuum:

- Towards the left (near physical reality). Users remain present in their physical environment and are never made to believe they have left that environment.
- In the middle (fully Mixed Reality). These experiences blend the real world and the digital world...

- Towards the right (near digital reality). Users experience a digital environment, and are unaware of what occurs in the physical environment around them. (Microsoft, 2020)

Hence, the mixed reality spectrum implies that shifting from one end of the spectrum to the other involves VR users being "made to believe" in the apparent worlds they experience. Accompanying the commitment to a distinctive ontology of VR objects then, is what we might call a *doxastic requirement*: users must believe in VR objects and worlds to perceive and interact with them. Again, while the Microsoft document is not a philosophical account, it does capture an attitude that is also present in the scientific and academic literature on VR. In Chapter Four I referred to recent work by Grabarczyk and Pokropski that characterizes our perception of VR worlds and objects as involving a kind of "ontological shift" (2016). They orientate their concern with an observation and a question:

> It is in fact deeply puzzling that virtual objects, environments and worlds can be so easily treated as "objects," "environments" and "worlds" respectively, and not merely as pictorial representations. When and why does a picture on a screen become a virtual object or place? One obvious intuition would be that it has something to do with the move from a 2D to a 3D plane, but there has to be more to it than this. When a camera pans an object in a movie, people still treat the movie as an animation and not a virtual world....
>
> (Grabarczyk and Pokropski, 2016: 36)

Immediately, this is vague about the relationship between 2D picture planes and 3D pictorial configurations. But more importantly, Grabarczyk and Pokropski do not give an account of how the apparent perception of objects, environments, and worlds might arise from pictorial perception in the case of VR. Instead, they rely on the unconvincing notion of an ontological shift to cover this theoretical gap, and quickly jump ahead to an interesting and mostly convincing account of how Gibsonian affordances might be used to explain spatial presence.

Charitably, their position might be considered a case of twofold seeing, because they seem to identify as part of the perceptual act involved in VR, the seeing of a 3D configuration in a 2D surface. But it quickly goes beyond this into the realms of ontological realism because they sign up both to the reality of the objects and worlds of VR, and a doxastic requirement that a user must believe that they encounter these things. Unless we take their language to be merely figurative, this is clear when Grabarczyk and Pokropski argue

> that the obvious prerequisite to being immersed or present in a virtual environment is to believe that there is some kind of alternative place in which we can immerse ourselves or be present (as opposed to a simple picture or an animation we can only look at from the outside). [...] Not only does the user have to believe in an explorable space, but also that the avatar she explores the space through remains embedded in this space as one of the objects in it.
>
> (2016: 35–36)

Here, the space, the objects within it, and the user's beliefs about them, are literal commitments of the account.

We need to take some care in interpreting Grabarczyk and Pokropski's doxastic requirements on VR, however, because viewing pictures, including VR pictures, *can* generate beliefs. First, it is the case that sometimes picture viewing does generate *mistaken* beliefs in the viewer that what is being viewed is a real space or object, and that the viewer bears a specific spatial relationship to it. This was the case in the mistaken perception of *trompe l'oeil* pictures discussed earlier, and in the case that I briefly mistook the picture of a street at Disneyland for an actual street. Second, beliefs are generated during non-mistaken picture perception, and these are typically beliefs about the picture's content: for example, that such a picture depicts a street receding away from the viewer. These need not be beliefs about actual things—though they may be if the picture is a documentation of reality—but rather, they are principally beliefs about the second fold of picture perception. Thus, it is reasonably clear that viewing pictures, and their instances in VR, involves beliefs, either mistaken beliefs, or more likely, unexceptionable beliefs about pictorial content.

But this is not what Grabarczyk and Pokropski take the doxastic requirement to be. Rather, it is their position that the viewer of VR simply believes themselves to be in an "alternative place." This is not a mistaken belief, because nothing else Grabarczyk and Pokropski say gives this impression, nor is it an unexceptionable belief about the pictorial content of VR, because they deny VR is a simple picturing process. Rather, the belief is part of an ontological shift in the VR user's doxastic attitudes. But it is simply not credible that users of VR have such beliefs or ontological commitments. Aside from the deceptive counterexamples of the swaying or defensive gestures noted earlier which I ascribed to a kind of sensorimotor illusion, and hence, a perceptual mistake, there is little behavioral evidence that users of VR take themselves to be in alternative places. *Pace* Slater, users of VR do not *really* run from virtual fires because they are all too aware of the virtual media context in which such fires are encountered. The complex behaviors that users of VR *do* exhibit—interacting with the game and making control inputs, perhaps to *fictionally* run from a fire—are evidence of their beliefs that they are using a VR system, and that they understand the interactive pictorial conventions of this medium.

Even the common trope of YouTube videos where an unsuspecting user is exposed to a virtual rollercoaster, and immediately begins screaming and flailing about, is not convincing evidence of beliefs in the reality of the experience. Rather, these responses, and the response I witnessed first-hand of myself screaming loudly when confronted by the *thing* in *Resident Evil VII*, are fundamentally likely to be evidence of the richness of the imaginative engagement in such cases. After all, readers of novels too, are capable of expressing real and rich affective responses such as crying, even when there is no evidence that they believe the things they read about have actually happened. The differences with the responses in VR derive from the egocentric interactive pictures of its medium, and subsequently, the kinds of imaginative engagement and responses this makes possible. The reason I was reluctant to play *Resident Evil VII* after that experience, was not because I believed I might be harmed in that alternative space, but that I knew about the rich game of make-believe it had in store for me when I put on the headset.

Moreover, the conjectured existence of these unlikely beliefs is redundant anyway, given that object recognition in pictures can function irrespective of any ontological commitment to the apparent objects pictured, even when those pictures are interactive. The perception of the apparent spatiality of Velasquez' *Las Meninas* does not depend on beliefs about the reality of the depicted room; and viewing a popup book similarly does not require beliefs about the reality of the dog running across the field. Of course, experiencing VR does not much feel like viewing *Las Meninas* or a popup book, but the reason that viewing VR does not feel like "a simple picture or an animation we can only look at from the outside" is because *VR is not a simple picture*, but a quite complex egocentric picture where a spatial impression is precisely conveyed, and where the interaction with the picture is itself pictorially rendered. Nevertheless, despite their egocentric and interactive nature, VR pictures, like traditional pictures, do involve the user looking at something from the "outside," if we take this rather figurative phrase to refer to the fact that all picture viewers principally engage with the surface of an image but see "through" this into a depicted space.

The lesson here is that this assertion of ontological commitments or beliefs misses an important step in the act of viewing VR media, that is, the act of perceiving the content inherent in the shapes and animations on the VR headset. Conveying this as an "ontological shift" is unhelpful because the process is not one of ontology at all, but of perception: if anything, there is a *perceptual shift* that occurs when one sees a virtual object, and this perceptual shift is the act of pictorial "seeing in." Ontological considerations only enter the picture, so to speak, once the specific use of the picture has been settled: that is, when it is decided whether the picture depicts an imaginary scene, a documentary one, or whether it transparently depicts something physically present to the user.

7.3 Digitalism about Virtual Objects

Another reason for taking VR to have a metaphysical significance begins with the question of precisely what is encountered in acts of VR seeing. In his recent work David Chalmers' ontological orientation on VR is clear from the very first sentence of the paper, when he asks, "How real is virtual reality?" (2017: 309). Additional questions that share this metaphysical focus quickly follow:

(1) Are virtual objects, such as the avatars and tools found in a typical virtual world, real or fictional?

(2) Do virtual events, such as a trek through a virtual world, really take place?

(309)

In his answer to these questions, a different conception of ontological realism about virtual objects becomes evident when Chalmers claims the objects depicted by VR headsets are "digital objects, constituted by computational processes on a computer" (2017: 317). Chalmers refers to this position as "digitalism," and contrasts it with the "virtual irrealism," which he takes to claim that virtual objects are unreal or fictional things (310). This analysis leads him to make some very strong claims about the ontological status of the apparent VR objects and worlds we encounter on computers, including, as we noted earlier, that "virtual reality is a sort of genuine reality, virtual objects are real objects, and what goes on in virtual reality is truly real" (309).

There is much to like in Chalmers' account, however, in many ways it runs together ideas that make it rather unhelpful for the explanation of VR media or their effects or use. This is not Chalmer's intention of course, as it is evident that he is rather more interested in squeezing some metaphysical juice out of virtual apples. In an earlier paper Chalmers discusses perfect virtual worlds (2003) and this more recent work somewhat follows in those initial steps. However, as is evident from the reference to *Second Life*—a virtual world that does not involve stereoscopic viewing—and the application of digitalism to even this case (Chalmers, 2017: 316), the focus is often on less than perfect virtual worlds also. This lack of differentiation between the cases is problematic because it is not always clear in Chalmers' paper whether the focus is virtual world apps and games, virtual reality media, or perfect virtual worlds. As we have found in this book, these are all very different things.

A first question is whether Chalmers' digitalist claims such as "virtual reality is a sort of genuine reality" really live up to their exciting billing. Whether the claim that virtual objects are "real" is credible, depends very much on precisely what we take the reference of the term to be. The digitalist position might amount to the relatively unexceptionable claim that there are depictive objects in the case of VR, or that there is some computational object—a 3D polygonal model perhaps—grounding our visual encounters with VR media. Both interpretations are consistent with the theory I have presented in this book. But Chalmers makes more ontologically adventurous claims that imply a stronger reading:

> The virtual world of *Second Life* involves virtual bodies (avatars) in virtual space. Virtual bodies are distinct from physical bodies, and virtual space is distinct from physical space. We really have these virtual bodies, as well as having physical bodies. There is nothing fictional about this.
>
> (Chalmers, 2017: 316)

So, it is not just that apparent virtual worlds, objects, and bodies depend for their depictive realization on graphical artefacts, or computer objects or processes, but that virtual worlds and bodies are *themselves* "digital objects, constituted by computational processes on a computer."[3] Moreover, there are such things as virtual bodies, and they are distinct from physical bodies, but also presumably, somewhat like real bodies for us to really "have" them. That such virtual bodies exist is obviously an ontological claim; and it is also just a little mysterious. So why would one take this view?

Chalmers has two key arguments for this position, the "causal" and the "perceptual" arguments (2017: 317–318). These arguments stand as Chalmers' principal justification for the identification of virtual objects and bodies with digital objects. However, because they do not acknowledge their basic assumptions about the intentional objects of virtual media, they ultimately fail in this goal. Both of the positive arguments are of a similar form, and I will deal only with the perceptual argument here. It runs like this:

(1) When using virtual reality, we perceive (only) virtual objects.
(2) The objects we perceive are the causal basis of our perceptual experiences.
(3) When using virtual reality, the causal basis of our perceptual experiences are digital objects.
(4) Virtual objects are digital objects.

(Chalmers, 2017: 318)

Formally, the argument is easy enough to understand. In terms of its content, Chalmers claims that the first premise is "intuitively plausible," that the second is uncontroversial in the philosophy of perception, and that the last is empirically justified. Thus, when you visually encounter your own body in a VR application—perhaps by looking down in *Half-Life: Alyx* to see your hands—what you perceive is a digital object, and, moreover, your virtual body *just is* this digital object.

However, the argument gets off to a bad start if we acknowledge the real variety of uses of virtual media. The plausibility that Chalmers ascribes to the claim that one perceives only virtual objects may derive from the trivial understanding that *any* object conveyed by virtual media is a virtual object for just this reason, just as any object conveyed by a picture is a pictorial object. But neither description has any ontological significance, and on closer inspection, because of the variety of uses of VR discussed in Chapter Five, where we found that there are multiple kinds of thing that the user can perceive through VR, it is false that we perceive only "virtual objects" in anything other than this trivial sense. The objects I have argued to be perceived when using virtual reality media include the picture surface, the apparent 3D objects configured in this surface, and the intentional object of the configurations, which might themselves be documentary or real objects and even objects really present to the user, and with which they really interact. Such was the case with the ball depicted in the Disney experiments; trivially this is a virtual object, in being the intentional object of a virtual depiction, but it is also an actual ball that the user perceives (and, which causes the user's visual experiences, though circuitously). Furthermore, it is possible that a given instance of VR depicts fictional acts of perception where the apparent intentional objects may not exist at all.[4] This variation undermines Chalmers' beginning assumption that there is a single and simple answer to his question of the identity of virtual objects.

In fact, Chalmers' ontological orientation on the nature of virtual objects and virtual seeing leads him to deny that some of these things are perceived. For example, in claiming that seeing in VR is more like "ordinary seeing" than seeing pictures, he denies that the immediate perceptual object is an image on a stereoscopic screen when he says that "in typical VR, one need have no sense of seeing a screen, and it can perhaps be argued that one does not see a screen at all" (Chalmers, 2017: 319). I have already argued here that VR users must see the screen—even if they are mistaken about what they see, or do not attend to it—because it is the screen that encodes the spatial configurations key to VR picturing. The other two aspects of VR seeing that for Chalmers make it more like ordinary seeing—the "three-dimensional perceptual experience from a perspective" and that one can "change one's perspective, and act on the world," are, I have argued, things that make VR a fascinating expansion of pictorial form, rather than something that disqualifies them from that category. Chalmers also gives no impression of acknowledging that the intentional objects, and causes, of virtual reality experiences might be real or documentary items, and that VR has prosthetic uses where "the causal basis of our perceptual experiences" are real things, even if their perception is algorithmically remediated.

Chalmers instead identifies the perceptual objects of VR with computational data structures: it is these digital objects that we perceive when we don a virtual reality headset. This is an unlikely claim, however, because the data structures of a computer program are abstract things that do not even present themselves to visual experience in a way that they could be seen, as Chalmers himself admits (2017: 322). Unlike the ball that I contend is seen in the Disney experiments, data structures lack a precise position such that they could bear the egocentric spatial information that philosophers such as Cohen and Meskin think is crucial to seeing (2004). Rather, what must be seen is *the effects* of these data structures on the configurations of a visual display, and, in some cases, the objects thus represented by these visual displays.

Moreover, data structures are not much like bodies or spaces, and as Chalmers also acknowledges, they only manifest sensorial qualities such as colors and sizes when seen through a stereoscopic viewer. As he notes, "A virtual flower may be red, while the corresponding digital object is not red," and that a virtual object is red only when it causes "reddish experiences in normal VR conditions," and that these normal conditions "currently involve access through an appropriate headset" (Chalmers, 2017: 22). But this is also problematic, because, first, it does not apply to virtual worlds such as *Second Life* that do not involve a stereoscopic headset, which Chalmers also includes in his virtual explanandum. Secondly, one simply wonders how this act of viewing occurs if regarding such a headset, it "can be argued that one does not see the screen at all." And finally, in some respects this theory is beginning to look suspiciously like a pictorial theory of VR, where one is seeing not digital items like data structures, but rather simply *pictures* (even if some of these may involve digital elements). That to see the properties of a virtual object "currently involve access through an appropriate headset" (22) is hence not the metaphysically exciting picture we are sold at the beginning of the account where "virtual reality is a sort of genuine reality," but rather a more sensible position that virtual objects and their properties are, most basically, the apparent objects and properties of pictures on a VR headset.

Curiously, in an argument that it is the digital objects themselves that are seen, Chalmers seems to agree with Walton's transparency thesis when he claims that "[i]t is widely accepted that when we look at a photograph or a film clip of Winston Churchill, we see Winston Churchill," then uses this observation to back his digitalist position that we really see digital objects (Chalmers, 2017: 319). Forgetting Chalmers' dubious claim about the philosophical consensus on the transparency thesis, surely the real lesson to be drawn from the example is that in some cases at least, viewers of virtual media see the objects depicted in virtual media rather than the data structures themselves, and that these data structures are only one (algorithmic) link in the chain that allows for VR pictorial seeing of cave paintings, zombies, New York City, and balls being thrown in your direction.

So, what should we make of Chalmers' claims that users have virtual bodies and see and interact with virtual spaces and that these are digital objects or data structures? I do not think these claims make all that much sense, and suspect that they owe to what is an over-complicated and confused account of VR.[5] Chalmers vacillates between cases and claims, and it makes his precise position hard to pin down. And again, like Grabarczyk and Pokropski's account, there is a gap in Chalmers' approach: that between computational data structures and the appearance of virtual objects. Filling that gap with a pictorial theory of VR simplifies things considerably, as one can easily see how such data structures could be used in producing a pictorial appearance of a body or space which can be viewed through a stereoscopic headset without the ontological implications Chalmers claims.

But such a pictorial theory as I have presented here does not mean that these VR pictures of bodies and spaces are causally impotent, and that they cannot meet the interactive and social functions that Chalmers' notes that VR is capable of—for example, "socializing, gathering information, or communicating with colleagues" (2017: 316)—because like other kinds of pictures, VR pictures are ripe for social uses; and pictures have always had this social potential over the deep history of their cultural existence. Moreover, VR pictures are often combined within "virtual suites," such that they may remediate and take on the complex social and informational functions of the items that they virtually picture, including the chat and socialization that might also occur on that now ancient virtual technology, the telephone. I will have more to say on this potential in the final section of this chapter.

Ultimately, Chalmers' ontological starting point and the resulting framing of his position in opposition to VR fictionalism means that he cannot fully acknowledge the compatibility of virtual realism and virtual fictionalism as being distinctive *uses* of very real VR media. Because of the perceptual and causal arguments, it does not make sense that one could perceive fictional things *through* virtual representations, because this would demand that one causally and perceptually encounters something that may not exist. Hence, what we must encounter in virtual reality is restricted to the digital object that is "directly encountered." But this undermines not only virtual fictionalism but also what I have referred to as virtual documentary and virtual transparency, that is, the idea that we might see real things through virtual representations. A pictorial conception of VR accommodates all these uses, and in a non-ontologically ambitious way, and so is to be preferred to Chalmers' digitalism.

7.4 The Social Reality of Virtual Worlds

Another way to explain the ontology of the apparent objects of virtual reality is to claim that they are in some sense social objects, perhaps in the sense of being "socially constructed." Philip Brey (2014) begins his discussion of the ontology of virtual objects with the observation that, "It is a common belief that objects in virtual environments are not real but are mere imitations or simulations of real objects," and follows this with the observation that, "Currently, there is widespread ontological confusion about virtual reality and its relation to the real world, which contributes to a flawed understanding of virtual reality and its potential" (2014: 43). As the reader might expect, I cannot help but agree with Brey's assessment, but I am also not sure that his own position avoids the confusion regarding VR. This is because, again, Brey's orientation at the outset is on the ontology of virtual objects and worlds, rather than the simpler framework of VR considered as a media form.

Brey argues that the confusion regarding virtual reality derives from the improper use of language. He claims that while most virtual objects have no physical existence, this does not mean that they have no existence at all. "Virtual apples," to take his example, "simulate or imitate real apples," and their ontology is one of being the "symbolic structures" that underlie computational processes (Brey, 2014: 43–44). Thus, like Chalmers, Brey takes virtual objects to be "digital objects." Moreover,

> Digital objects qualify as objects because they are persistent, unified, stable structures with attributes and relations to other objects, and agents can use and interact with them. [...] A *virtual object* is a digital object that is represented graphically as an object or a region in a two- or three-dimensional space and that can be interacted with or used through a computer interface.

(44)

And like Grabarczyk and Pokropski, Brey argues that the perception of virtual objects has a doxastic requirement, though this is framed not as the formation of beliefs, but as the "suspension of disbelief": "just like immersing oneself in a movie requires one to experience or perceive depicted events as if they are actually happening, immersion in a virtual world requires one to act as if it is real" (Brey, 2014: 45). Many virtual reality objects for Brey—rocks, apples, and other physical things—are thus merely fictional, because even though they are simulated, they have no real existence.

But drawing on John Searle's famous account of social reality (Searle, 1995) Brey argues that at least *some* virtual objects can genuinely be realized in virtual reality because of their institutional ontology: "Physical reality and ordinary social reality can usually only be simulated in virtual environments, whereas institutional reality can in large part be *ontologically reproduced* in virtual environments" (2014: 47, emphasis in original). Physical objects can only be simulated because the virtual reality is not capable of embodying their characteristic qualities (that is, their physical extension, mass, and so on), whereas virtual reality *is* capable of remediating the characteristic features of objects with an institutional ontology such as money (that is, fungibility, discreteness, scarcity, and so on).

In many ways, Brey's account is consistent with the media account of VR I have presented here because it grasps the key fact that virtual objects such as apples are represented by computational media and are often fictional. And I agree that some previously physical items like currency can be *virtualized* because they have an institutional ontology. The account has problems, however. First, as an account of virtual reality media and its apparent objects, Brey makes a number of implausible claims about what such virtual objects might be and how they might reproduce the features of "real" objects. And second, like the other accounts of VR considered in this chapter, Brey conflates stereoscopic virtual reality media with computational virtualization more generally, shifting quickly between the two quite distinct phenomena. His account is more successful for explaining the latter, but it is severely compromised by having been framed as an explanation of the former.

First, then, Brey provides an implausible account of VR objects. Particularly problematic is that the "digital object[s] that [are] represented graphically as an object or a region in a two- or three-dimensional space," that Brey contends to constitute virtual objects such as apples and circles, simply *are not* "persistent, unified, stable structures," in a way that would allow that identification. First, virtual depictive items are not persistent. *Occlusion culling* is the technique where surfaces of a graphical model that cannot be seen from the virtual perspective of a user are not rendered by the graphics hardware (LaValle, 2019: 199). This is desirable because rendering objects that cannot be seen is essentially a waste of processing power that could be utilized in other areas of game performance. Thus, aspects of a digital model that cannot be seen from a given perspective, and the objects they compose, are not persistent, but come into and out of existence as they are revealed by the virtual perspective. And so, object permanence is not a safe assumption in virtual worlds!

Second, the virtual depictive artifacts underlying an apparent virtual object such as an apple need not be unified. This is because a single apparent virtual object in a VR game might have different graphical representations under different contexts of use. This is quite common in videogames where single items encountered in the fully 3D world might be depicted by graphical artefacts different to those used to represent the item when it sits in the player's inventory. In the survival game *DayZ* (2018), an apple picked from the ground under a tree is depicted by a different graphical artifact than the same apple when it is in your bag. Apparent virtual objects are not unified, rather, the program produces graphics that give the impression of unified objects: in the case of *DayZ*, fictional objects such as apples one might pick up, store in a bag, and retrieve later to eat. Moreover, *any* apple encountered in the game will be depicted by the same set of digital assets, resisting the identification of any specific apple with a specific digital object. This is also the case in VR games such as *Resident Evil VII* and many other VR applications.

Finally, virtual depictive artifacts are not stable, as digital graphical objects have a habit of blinking in and out of existence due to glitches. *DayZ* itself is notoriously "buggy" in this regard, with large parts of the world sometimes flickering in and out of existence around you. Hence, digital artefacts are not persistent, unified, and stable in the way that *apparent* virtual objects are, and in fact, there is likely no single unique digital structure corresponding to such virtual objects at all. To reemphasize the central thesis of this book, VR is a picturing device that may give the impression of encountering persistent, unified, and stable objects in one's egocentric space, but whether the objects depicted have such features is itself determined by the intentional context of use of such a pictorial medium.

Brey's account of how virtual reality media might *reproduce* the objects they depict is also not always convincing. Virtual objects are considered capable of counting as real objects when they reproduce the characteristic features of those objects. After claiming that a virtual performance of Bach's *Toccata and Fugue in D Minor* might count as a real performance (a claim that glides obliviously over the possibility that musical performances might themselves have ontological requirements in terms of the *means* of performance) Brey notes that

> Similarly, when in a virtual environment a circle is drawn, the result is a real circle, since a circle is mathematically defined as a phenomenon consisting of points on a plane, and is not by definition a physical object with weight and mass.
>
> (2014: 46)

However, even though a virtual representation of a circle might on some occasions produce a circle on the screen of the stereoscopic headset, it need not. What it is likely to produce is what we earlier referred to as an "occlusion shape," that is, a shape which when viewed from an apparent perspective gives the impression of constituting a circle (Hyman, 2006: 81). Often (if not usually) these occlusion shapes will be ellipses because the virtual perspective on the circle will be from "side on." Some virtual circles may never reproduce a real circle, for example, if it is only ever possible in the virtual environment to view them obliquely. (And it is no solution to say that the digitally coded geometrical shape reproduces a circle, because it is even less circle-like than the occlusion shape, because it does not even count as any kind of shape.) This shows that it is wrong to think of virtual phenomena like circles as *objects* that are *reproduced* in a virtual way; it is rather, that virtual reality media are used to produce surfaces in which 3D configurations can be seen via pictorial seeing, and which can be treated as if they were objects for specified interactive purposes.

Brey's account of the reality of the social or institutional practices that can really be carried out by employing virtual media is more compelling. The features of virtual media that allow for such functions are indisputably real and they reproduce in a virtual way the functions and structures of customary such items: pieces of code are discrete and fungible, and this is why they can serve the function of currency; virtual world apps like *Second Life* allow for genuine conversations because they involve real text and speech chat. But these claims do not follow from, or necessitate, the account of the ontology of virtual reality objects that Brey provides. Indeed, it seems a change in topic to the more general issue of virtualization that was discussed in Chapter Two. This is a very different phenomenon to the graphical rendering of apples on VR headsets, in that the kinds of virtual institutional practices Brey considers are often not graphically rendered at all—this is typically the case with virtual currency such as cryptocurrencies. As such, that social and institutional practices can employ virtual media is independent of the question of the ontology of apparent VR objects.

The major problem here is the framing of these two issues—virtual reality media, and the virtualization of social practices—under the single banner of ontology. For Brey, virtual reality objects are to be approached such that it is coherent to ask if they are *real* such objects, and that the reality of these objects is assumed to be the key focus of the philosopher when she comes to explain virtuality. Let us assume that the account of VR I have presented here is correct: I think we can safely conclude that this orientation would certainly be an odd way to approach the explanation of other kinds of pictures. To claim, in parallel with Brey's beginning observation, that "it is a common belief that objects in pictures are not real but are mere imitations or simulations of real objects," would generate the obvious response: "Well of course that's not a pipe, it's a picture of a pipe!" But there is also a less obvious response: "But aren't some pictures of pipes, pictures of *real* pipes? And can't we use pictures of pipes to achieve certain real aims—investigating pipes, evaluating their aesthetic qualities, illustrating an advertising sign for a tobacconist—purposes for which a picture might serve just as well (and maybe better) than an actual pipe?" Taking a similar approach to VR picturing—questioning its intentional objects, and how such pictures may function in practical contexts—seems a far more profitable approach than pointing at a VR apple, pipe, or dollar, and asking whether it is real.

7.5 An Error Theory for Virtual Reality

In the critique above, I have repeatedly identified as a key fault in ontological realism the conflation of the medium of VR and its modes of depiction, with its apparent intentional objects. A closely allied confusion, and one which is equally mischievous, is that VR comprises not the remediation of spatial experience, but rather of objects and worlds thus experienced, and that this produces a distinctive kind of "virtual world" or "virtual object." For Grabarczyk and Pokropski, in virtual reality "Pictures become objects" (2016: 36). For Chalmers, "virtual objects are real objects" (2017: 309). And for Brey, "virtual apples simulate or imitate real apples" (2014: 43). So why might one claim that VR amounts to the remediation of objects and worlds rather than just the apparent experience of these in a distinctive, perhaps realistic, picturing medium? Explaining this will allow for somewhat of an *error theory* for why ontology has so captivated theorists of VR.

Of course, part of the reason is the expression "virtual reality" itself. We might judge that it is the linguistic predominance of this phrase that is partly at fault for sending the philosophy of virtuality down the metaphysical rabbit hole. This strange and seemingly oxymoronic phrase is so catchy, so prominent in common discourse, and so conceptually fecund for theorists and philosophers with a metaphysical bent, that it almost always becomes the focus of discussion. This is unfortunate because it may be a case where a piece of fashionable terminology has overdetermined a theoretical issue and potentially excluded other ways of thinking about the phenomenon. The term "virtual reality" itself is relatively recent and though it is often attributed to the technologist, scientist, writer, and artist Jaron Lanier, who in the 1980s was one of the first people to work on commercialized VR technology (Kahn, 2011) it has an older precedent in Antonin Artaud's notion of "la réalité virtuelle" in his collection of essays, *Le Théâtre et son double* (*The Theatre and its Double*, 1938/1958). But VR media, as I have argued here, predate even this reference in the sense that Alberti's pictorial techniques might be considered as a kind of virtual media because of their intention to remediate the experience of space. Ultimately, "virtual reality" is just one terminological entry point into the broad set of issues considered here, and we should not let it dominate our basic conceptualization of these phenomena.

But there is also something else here, and this is the persisting conflation of virtualization—the remediation of items in a non-actual and unfamiliar medium, often computational in form—with VR media specifically, which are themselves a special case of virtualization amounting to the remediation of native experience through egocentric interactive pictures. Aiding and abetting this conflation is the continual running together of examples of virtual worlds and virtual reality media in the literature, a problem that afflicts each of the sources considered in the earlier parts of this chapter. These phenomena are conflated, I believe, because in practice they *are* sometimes combined: that is, *a virtualized object or activity might also be the intentional object of an egocentric interactive picture.*

For example, a virtualized phone might itself be depicted in VR media. Earlier in this book I referred to the PS4 VR game *The London Heist*. This game contains several VR vignettes that give an impression of the capabilities of PS4 VR, acting as a taster or a demo of the technology. In the example discussed in an earlier chapter, we might reach out to be handed a mobile phone, answering it. Now let us change the example a little: imagine that when you answer the phone, the call is not from a fictional criminal, but from a real friend, because your own actual phone has been given a visual and interactive rendering by the VR media of the game. Imagine then that this virtual phone has the full functional efficiency of an actual phone, and all these functions can be accessed within the VR environment by interacting with controllers and making tracked gestures. Given current technology, this is not an unlikely thought experiment.

It might be objected: should we not say that in this instance that VR has remediated the phone itself? I am not sure we should. For sure, this would be an instance of a virtualized phone, and one would encounter it in an apparent virtual reality. But the virtualization of the phone here owes to the more basic sense of virtualization, that is, the functional efficiency of an item being instantiated in a non-customary way. This is the sense in which we encounter a phone via an icon on our computer desktop; and a mobile phone on an Android device itself counts as a virtual remediation of a phone. But in neither of these cases do we encounter the phone through VR media (that is, via an egocentric interactive picture). And most of the time that we encounter such virtual items it has nothing to do with VR media at all. It turns out though, that it could do, in that VR media might be employed to remediate our perceptual access to such virtualized items.

What can be seen here is the potential for the *nesting* of virtualized items and functions within the experiential spaces of VR media. VR media, in this way, typically comprise *suites* of virtual objects, processes, and functions. A virtual media suite, in the sense I intend here, is a set of algorithms, functions, programs, hardware, and peripherals that are combined into a VR application to allow for a given VR experience. *Suite* refers to the compartmentalization of many of the aspects of VR media, and that they might be combined in various ways to allow for different VR experiences and activities. *The London Heist* involves a virtual suite that gives the experience of having a dangerous, exciting, and somewhat frightening criminal adventure. VR *Rec Room*, is a cross platform VR suite that allows for users to chat with other real people, play games, and construct their own virtual spaces. This reference to a suite emphasizes the distributed and often modular nature of the functionality of VR applications, and this kind of functional compartmentalization is basic to computing technology when we consider that the tasks conducted by computers decompose into smaller and smaller units—particularly, programs, algorithms, and code. Here though, compartmentalization is considered in a broader often physical sense, in which peripherals such as stereoscopic headsets might be considered as discrete or substitutable parts of a virtual suite.

These virtual media suites, however, because of their functional coordination, may give the impression that the functions and items they depict exist as singular and integrated entities that are simply remediated instances of their corresponding actual-world versions. A virtual suite and its wide functionality, and that these are presented to us via VR experiential media, might give the impression of *virtual worlds* or *objects* themselves being remediated in a "persistent, unified, and stable way." The VR mediated cell phone hypothesized earlier may function as a phone simpliciter by allowing for placing phone calls, browsing the internet, sending texts and so on. But though when depicted by VR media this virtual phone might give the superficial appearance of inhering in a single object, it is in fact a suite of virtual media functions: a 3D VR model, motion-tracking allowing one to grasp and move the phone, a set of sounds, a vibrate function in the haptic controller, and a network connection allowing for telephone calls and internet browsing. In a non-virtual phone, these appearances and functionality inhere in a single physical object. But when depicted in the egocentric interactive pictures of VR, they devolve to a suite of functional and depictive virtual assets that give the impression or experience that one is simply dealing with a phone. Moreover, it is not the visually encountered interactive graphical model that is the virtual phone, but the whole suite. This observation, I believe, deflates much of the intuition that VR media simply remediates objects with a virtual existence. It also means that trusting the phenomenology of VR is not a reliable guide to discerning the actual nature of the phenomenon.

A similarly hypothetical case of VR conferencing apps provides another example to bear out the point of why it is not the apparent objects of our perception that are being remediated by VR, and how such virtual suites might expand to encompass customary social functions. During the period of the COVID-19 pandemic and associated lockdowns, many of us adapted to the experience of holding meetings or teaching via video conferencing apps such as Zoom or Skype. In late 2020 I attended the annual meeting of the American Society for Aesthetics in this way, where I spoke about virtual picturing to a distanced but virtually present audience. I have to admit that while I appreciate that these virtual apps made possible a certain amount of normal business during this unfortunate time, I was never entirely comfortable with these virtualized meetings or talks. A small 2D picture of a distant muted colleague or friend was scant replacement for their actual presence. The limitations of the apps led to some social discomfort, as I felt, speaking in my office at 3am New Zealand time to an audience scattered around the world, that my words were being cast into the void. In my experience, to give an effective philosophical talk, or at least to feel that I have done so, I need to see the eyes and gestures of my audience; to see if they smile, laugh, or lean into our conversation. It is not just that video conference apps do not allow us to perceive these things; they may inhibit the possibility of their occurrence.

I sometimes wondered whether the utilization of a VR medium might have improved matters, that is, where the participants, wearing VR headsets, might "meet" in VR and thus have a kind of bodily presence with others in a virtualized space. Something like this is already available in the VR chat spaces mentioned above, but a dedicated VR conference suite might add additional functionality specific to conference presentations. This, I speculated, might add to a sense of connection between the participants. Such an app would not remediate the participants in any sense—they would remain as they are, people distant from me, merely meeting virtually via an app—but what it would do is remediate my experiential and interactive access to those other people, to their responses and their gestures, and to their copresence as a group around me. It would give a sense of them feeling *present* to me and me to them. But it is again clear in this case that the intentional objects of the VR app are the people meeting, and they retain their real existence.

Finally, let us indulge the metaphysician of virtual reality and ask the question of the reality of the objects depicted in virtual media. Are the objects, worlds, people, and activities depicted in VR real? This is clearly an ambiguous question, and its answer very much depends on which aspects and uses of VR we mean to refer to. In the context of VR surfaces, these are real, and we see them. The apparent 3D configurations seen through these VR media are real, and it really is the case that our visual systems perceive the surface as an *apparent* 3D configuration. Many of the items depicted by the 3D configurations in documentary VR—New York City and the paintings in the La Garma cave complex—are real even if they are not spatially or temporally present to the user. Many of the objects and characters of videogames are unreal in comprising fictional characters, places, and beasts. But some of the inhabitants and places of videogames are also more like documentary objects, as is the case with non-VR fictions also. The ball that is thrown to the user and which they catch is certainly real. By inspecting actual VR media and their wide variety of uses, we can see that all these judgments can be made about the intentional objects of VR pictures, and also, that decontextualizing and collapsing together the varied uses to which VR is put is likely to result in the confusion that there is something like a unified class of "VR objects."

All of this interesting detail is obscured by the thought experiment of perfect virtual worlds. Ultimately, we do not stand to learn much about virtual media by employing this thought experiment. Indeed, perfect virtual worlds, and the questions and intuitions they arouse, are precisely the wrong place to start in a study of VR media. No doubt there are metaphysical questions prompted by the consideration of perfect virtual worlds, but these already pertain to other forms of Cartesian fictions and adding the issue of virtual worlds to the debate adds little value beyond providing a modern framing for these issues. In fact, this framing does harm by adding confusion to the issues with VR media. It is always worthwhile to frame philosophical issues in ways that motivate those issues for readers and students, of course, but in this case, doing so will mean that the real interest in the *aesthetics* of VR media is obscured.

The result of the analysis presented in this book is that virtual media are an *extension* of aesthetic technologies, and not a revolution. They are precedented in previous media and much of their hype derives from investing them with a technical or metaphysical significance they need not have. Nevertheless, VR media are of interest in that the incremental development of depictive technology they comprise adds to our understanding of depiction and of aesthetics itself. Deflating the metaphysical temptations hence does not mean that VR media are uninteresting. In fact, their interest is quite deep, as I hope to have shown throughout this book. They allow for many interesting discoveries about aesthetics, both in the sense relating to art and the aesthetics of natural items, but also in the deeper sense relating to our perceptual and agential access to the world.

Notes

1 See for example the collection of essays in Mark Silcox's *Experience Machines* (2017).
2 A 2019 issue of the philosophical journal *Disputatio* is devoted to addressing Chalmer's claims about VR objects, and every paper is orientated, in one way or another, on these ontological issues. VR media themselves, hardly get a look in.
3 McDonnell and Wildman (2019) also identify this strong reading of Chalmers' claim, and their paper is a helpful and detailed critique of Chalmers' views on VR, even though I do not agree with the largely fictionalist position they ultimately develop.
4 Obviously, there is some philosophical work to do here to explain how there can be non-existent referents in the case of fictions, but I do not see it as my problem to solve, as it is a quite general problem with accounting for the apparent ontological commitments of fictional statements. With respect to the apparent "bodies" depicted by fictions, I think there are promising accounts available (Brock, 2002).
5 I am not the only philosopher to think that Chalmers' position is unclear; see McDonnell and Wildman, 2019: 4–5.

Bibliography

Aarseth, E. (2007) "Doors and perception: Fiction vs. simulation in games," *History and Theory of the Arts, Literature and Technologies* (9): 35–44.

Abell, C. (2006) "Realism and the riddle of style," *Contemporary Aesthetics*, 4: 376.

Alberti, L. B. (1435/1973) *On Painting*. Translated by J. R. Spencer, New Haven: Yale University Press.

Al-Khalili, J. (2015) "In retrospect: *Book of Optics*," *Nature*, 518: 164–165. https://doi.org/10.1038/518164a.

Anable, A. (2018) *Playing with Feelings*. Minneapolis: University of Minnesota Press.

Anderson, P. L., Price, M., Edwards, S. M., Obasaju, M. A., Schmertz, S. K., Zimand, E., & Calamaras, M. R. (2013) "Virtual reality exposure therapy for social anxiety disorder: A randomized controlled trial," *Journal of Consulting and Clinical Psychology*, 81 (5): 751–760.

Aristotle. (1984) *The Complete Works of Aristotle: The Revised Oxford Translation*, Volume II, edited by J. Barnes. Princeton, N.J.: Princeton University Press.

Artaud, A. (1938/1958) *The Theatre and its Double*. Translated by M. C. Richards. New York: Grove Weidenfeld.

Aulisio, M. C., Han, D. Y., & Glueck, A. C. (2020) "Virtual reality gaming as a neurorehabilitation tool for brain injuries in adults: A systematic review," *Brain Injury*, 34 (10): 1322–1330.

Bacon, F. (1605/2013) *The Advancement of Learning*. Cambridge: Cambridge University Press.

Bartel, C. (2018) "Ontology and transmedial games," in *The Aesthetics of Videogames*, edited by J. Robson and G. Tavinor. New York: Routledge.

Bazin, A. (2009) "Ontology of the photographic image," in *What Is Cinema?* Translated by T. Barnard. Montreal: Caboose Books.

Bideau, B., Kulpa, R., Vignais, N., Brault, S., Multon, F., & Craig, C. (2009) "Using virtual reality to analyze sports performance," *IEEE Computer Graphics and Applications*, 30(2): 14–21.

Bostrom, N. (2003) "Are you living in a computer simulation?" *Philosophical Quarterly*, 53 (211): 243–255.

Brey, P. (2014) "The physical and social reality of virtual worlds," in M. Grimshaw (ed.), *The Oxford Handbook of Virtuality*. Oxford: Oxford University Press.

Brock, S. (2002) "Fictionalism about fictional characters," *Noûs*, 36: 1–21.

Burleigh, A., Pepperell R., & Ruta, N. (2018) "Natural perspective: Mapping visual space with art and science," *Vision*, 2 (2).

Cano Porras, D., Zeilig, G., Doniger, G. M., Bahat, Y., Inzelberg, R., & Plotnik, M. (2020) "Seeing gravity: Gait adaptations to visual and physical inclines. A virtual reality study," *Frontiers in Neuroscience*, 13. https://doi.org/10.3389/fnins.2019.01308.

Carrozzino, M., & Bergamasco, M. (2010) "Beyond virtual museums: Experiencing immersive virtual reality in real museums," *Journal of Cultural Heritage*, 11(4): 452–458. https://doi.org/10.1016/j.culher.2010.04.001.

Chalmers, D. (2003) "The Matrix as metaphysics." In *Philosophers Explore the Matrix*, edited by C. Grau. Oxford, UK: Oxford University Press.

Chalmers, D. (2017) "The virtual and the real," *Disputatio*, 9 (46): 309–352.

Cheah, C. S. L., Barman, S., Vu, K. T. T., Jung, S. E., Mandalapu, V., Masterson, T. D., Zuber, R. J., Boot, L., & Gong, J. (2020). "Validation of a Virtual Reality Buffet environment to assess food selection processes among emerging adults," *Appetite*, 153 [104741]. https://doi.org/10.1016/j.appet.2020.104741.

Chun, M. M., & Wolfe, J. M. (2001) "Visual attention," in *Blackwell Handbook of Sensation and Perception*, edited by E. B. Goldstein. Malden, MA: Blackwell.

Clay, V., König, P., & Koenig, S. (2019) "Eye tracking in virtual reality," *Journal of Eye Movement Research*, 12.

Cohen, J., & Meskin, A. (2004) "On the epistemic value of photographs," *The Journal of Aesthetics and Art Criticism*, 62 (2): 197–210.

Collewijn, H., & Erkelens, C.J. (1990) "Binocular eye movements and the perception of depth," in E. Kowler (ed.) *Eye Movements and Their Role in Visual and Cognitive Processes*. Amsterdam: Elsevier Science: 213–261.

Copeland, J. (2005) *Alan Turing's Automatic Computing Engine: The Master Codebreaker's Struggle to Build the Modern Computer*. Oxford: Oxford University Press.

Crane, T. (2013) *The Objects of Thought*. Oxford: Oxford University Press.

Croci, D. M., Guzman, R., Netzer, C., Mariani, L., Schaeren, S., Cattin, P. C., & Jost, G. F. (2020) "Novel patient-specific 3D-virtual reality visualisation software (SpectoVR) for the planning of spine surgery: A case series of eight patients," *BMJ Innovations*, 6(4).

Csikszentmihalyi, M. (1990) *Flow: The Psychology of Optimal Experience*. New York: Harper and Row.

DeBose, K. (2020) "Virtual anatomy: Expanding veterinary student learning," *Journal of the Medical Library Association: JMLA*, 108 (4): 647.

de la Peña, N., Weil, P., Llobera, J., Giannopoulos, E., Pomés, A., Spanlang, B., Friedman, D. V., Sanchez-Vives, M., & Slater, M. (2010) "Immersive journalism: Immersive virtual reality for the first-person experience of news," *Presence: Teleoperators and Virtual Environments*, 19 (4): 291–301.

Deleuze, G. (2002) "The actual and the virtual," In *Dialogues* II. Translated by E. Ross Albert. London: Continuum.

Descartes, R. (1641/1984) *The Philosophical Writings of Descartes*, Volume II. Translated by J. Cottingham, R. Stoothoff, D. Murdoch, & A. Kenn. Cambridge: Cambridge University Press.

Dilworth, J. (2010) "Realistic virtual reality and perception," *Philosophical Psychology*, 23: 23–42.

Dockx, K., Bekkers, E. M. J., Van den Bergh, V., Ginis, P., Rochester, L., Hausdorff, J. M., Mirelman, A., & Nieuwboer, A. (2016) "Virtual reality for rehabilitation in Parkinson's disease," *Cochrane Database of Systematic Reviews* (12). doi:10.1002/14651858.CD010760.pub2.

Edgerton, S. Y. (1973) "Brunelleschi's first perspective picture," *Arte Lombarda*, 18 (38/39): 172–195.

Edgerton, S. Y. (2009) *The Mirror, the Window & the Telescope: How Renaissance Linear Perspective Changed Our Vision of the Universe*. Ithaca, NY: Cornell University Press.

Ermi, L., & Mäyrä, F. (2005) "Fundamental components of the gameplay experience: Analysing immersion," *Proceedings of the DiGRA Conference on Changing Views: Worlds in Play, Vancouver*, 2005: 15–27.

Feagin, S. (1998) "Presentation and representation," *The Journal of Aesthetics and Art Criticism*, 56 (3): 234–240.

Fodor, J. A. (1975) *The Language of Thought*. New York: Thomas Y. Crowell.

Fossataro, C., Tieri, G., Grollero, D., Bruno, V., & Garbarini, F. (2020) "Hand blink reflex in virtual reality: The role of vision and proprioception in modulating defensive responses," *European Journal of Neurosciences*, 51 (3): 937–951.

French, E. (2010) Interview with Noam Chomsky: "Direct participation in creativity," retrieved 11/11/2020 from https://usa.anarchistlibraries.net/library/eric-french-interview-with-noam-chomsky-direct-participation-in-creativity.

Gabriel, R. A. (2006) *Soldier's Lives though History: The Ancient World*. Westport, CT: Greenwood Press.

Garner, T. (2018) *Echoes of Other Worlds: Sound in Virtual Reality Past, Present and Future*. Cham, Switzerland: Palgrave Macmillan.

Gaut, B. (2010) *A Philosophy of Cinematic Art*. Cambridge: Cambridge University Press.

Gibson, J. (1986) *The Ecological Approach to Visual Perception*. London: Lawrence Erlbaum Associates.

Gombrich, E. H. (1960) *Art and Illusion: A Study in the Psychology of Pictorial Representation*. Princeton: Princeton University Press.

Gombrich, E. H. (1987) "*Western art and the perception of space*," Space in European Art, Council of Europe Exhibition, Japan, 1987: 5–12.

Gombrich, E. H. (2006) *The Story of Art*. London: Phaidon Press.

Goodman, N. (1976) *Languages of Art: An Approach to a Theory of Symbols*. 2nd ed. Indianapolis, IN: Hackett.

Grabarczyk, P., & Pokropski, M. (2016) "Perception of affordances and experience of presence in virtual reality," *Avant*, VII (7): 25–44.

Grau, O. (2003) *Virtual Art: From Illusion to Immersion*. Cambridge, MA: MIT Press.

Gualeni, S., & Vella, D. (2020) *Virtual Existentialism: Meaning and Subjectivity in Virtual Worlds*. London: Palgrave Pivot.

Halliday, J. (1993) "Air operations in Korea: The Soviet side of the story," in *A Revolutionary War: Korea and the Transformation of the Postwar World*, edited by W. J. William. Chicago, IL: Imprint Publications.

Harman, G. (1973) *Thought*. Princeton, NJ: Princeton University Press.

Heesy, C. P. (2009) "Seeing in stereo: The ecology and evolution of primate binocular vision and stereopsis," *Evolutionary Anthropology*, 18: 21–35.

Heim, M. (1993) *The Metaphysics of Virtual Reality*. Oxford: Oxford University Press.

Heim, M. (1998) *Virtual Realism*. Oxford: Oxford University Press.

Hoffman, D. M., Girshick, A. R., Akeley, K. & Banks, M. S. (2008) "Vergence-accommodation conflicts hinder visual performance and cause visual fatigue," *Journal of Vision*, 8 (3): 1–30.

Hornsey, R. L., Hibbard, P. B., & Scarfe, P. (2020) "Size and shape constancy in consumer virtual reality," *Behavior Research Methods*, 52 (4): 1587–1598.

Hyman, J. (2006) *The Objective Eye: Color, Form, and Reality in the Theory of Art*. Chicago: University of Chicago Press.

Indovina, P., Barone, D., Gallo, L., Chirico, A., De Pietro, G., Giordano, A. (2018) "Virtual reality as a distraction intervention to relieve pain and distress during medical procedures: A comprehensive literature review," *Clinical Journal of Pain*, 34 (9): 858–877.

Jiang, M. Y. W., Upton, E., & Newby, J. M. (2020) "A randomised wait-list controlled pilot trial of one-session virtual reality exposure therapy for blood-injection-injury phobias," *Journal of Affective Disorders*, 276: 636–645.

Juul, J. (2019). "Virtual reality: Fictional all the way down (and that's OK)," *Disputatio*, 11(55): 1–11.

Kahn, J. (2011) "Visionary," *The New Yorker*, July 11/18, 2011 Issue, retrieved 17/11/2020 from https://www.newyorker.com/magazine/2011/07/11/the-visionary.

Kardong-Edgren, S., Farra, S. L., Alinier, G., & Young, H. M. (2019) "A call to unify definitions of virtual reality," *Clinical Simulation in Nursing*, 31(C): 28–34.

Klee, P. (1953) *Pedagogical Sketchbook*. New York: Frederick A. Praeger.

Klevjer, R. (2017) "Virtuality and depiction in video game representation," *Games and Culture*, 14 (7–8): 724–741.

Kubovy, M. (1986) *The Psychology of Perspective and Renaissance Art*. Cambridge: Cambridge University Press.

Kulvicki, J. (2006) *On Images*. Oxford: Blackwell.

Laver, K. E., Lange, B., George, S., Deutsch, J. E., Saposnik, G., & Crotty, M. (2017) "Virtual reality for stroke rehabilitation," *Cochrane Database of Systematic Reviews* (11).

Lawson, B. D. (2014) "Motion sickness symptomatology and origins," in *Handbook of Virtual Environments: Design, Implementation, and Applications*. Boca Raton, Florida: CRC Press.

LaValle, S. (2019) *Virtual Reality*. Retrieved 22/03/2021 from http://vr.cs.ui.uc.edu/.

Levinson, J. (1998) "On Wollheim," *The Journal of Aesthetics and Art Criticism*, 56 (3): 227–233.

Lopes, D. M. (1996) *Understanding Pictures*. Oxford: Oxford University Press.

Lopes, D. M. (2006) "The Special and General Theory of Realism: Reply to Abell, Armstrong and McMahon," *Contemporary Aesthetics*, 4. Retrieved 24/03/2021 from https://www.contempaesthetics.org/newvolume/pages/article.php?articleID=373.

Lopes, D. M. (2010) *A Philosophy of Computer Art*. Abingdon and New York: Routledge.

Ludlow, P. (2017) "Cypher's choices: The variety and reality of virtual experiences," in *Experience Machines: The Philosophy of Virtual Worlds*, edited by M. Silcox. London: Rowman and Littlefield.

Ludlow, P. (2019) "The social furniture of virtual worlds," *Disputatio*, 11(55): 1–25.

Maister, L., Slater, M., Sanchez-Vives, M. V., & Tsakiris, M. (2015) "Changing bodies changes minds: Owning another body affects social cognition," *Trends in Cognitive Sciences*, 19 (1): 6–12.

Manetti, A. D. T. (1970) *The Life of Brunelleschi*. University Park: Pennsylvania State University Press.

Matravers, D. (2014) *Fiction and Narrative*. Oxford: Oxford University Press.

Matthen, M. (2005) *Seeing, Doing, and Knowing: A Philosophical Theory of Sense Perception.* Oxford: Oxford University Press.

Mazyn. L. I. N., Lenoir, M., Montagne. G., & Savelsbergh, G. J. P. (2004) "The contribution of stereo vision to one-handed catching," *Experimental Brain Research*, 157: 383–390.

McDonnell, N., & Wildman, N. (2019) "Virtual reality: Digital or fictional?" *Disputatio*, 11 (55): 1–27.

Meehan, M., Insko, B., Whitton, M., & BrooksJr, F. P. (2002) "Physiological measures of presence in stressful virtual environments," *ACM Transactions on Graphics (tog)*, 21 (3): 645–652.

Microsoft. (2020) "What is Mixed Reality?" Archived at https://docs.microsoft.com/en-us/windows/mixed-reality/discover/mixed-reality.

Milgram, P., & Kishino, F. (1994) "A taxonomy of mixed reality visual displays," *IEICE Transactions on Information and Systems*, 77: 1321–1329.

Mooradian, N. (2006) "Virtual reality, ontology, and value," *Metaphilosophy*, 37 (5): 673–690.

Murphy, D. (2017) *"Virtual reality is 'finally here': A qualitative exploration of formal determinants of player experience in VR,"* Proceedings of DiGRA 2017, Melbourne, 2017.

Murray, J. H. (2020). "Virtual/reality: how to tell the difference," *Journal of Visual Culture*, 19 (1): 11–27.

Nanay, B. (2004) "Taking twofoldness seriously: Walton on imagination and depiction," *Journal of Aesthetics and Art Criticism*, 62 (3): 285–289.

Nanay, B. (2008) *"Picture perception and the two visual subsystems,"* in Proceedings of the 30th Annual Conference of the Cognitive Science Society (CogSci 2008), edited by B. C. Love, K. McRae, & V. M. Sloutsky, pp. 975–980. Hillsdale, NJ: Lawrence Erlbaum.

Nanay, B. (2015) "Trompe l'oeil and the dorsal/ventral account of picture perception," *Review of Philosophical Psychology*, 6: 181–197.

Nanay, B. (2018) "Threefoldness," *Philosophical Studies*, 175: 163–182.

Nash, K. (2018) "Virtually real: Exploring VR documentary," *Studies in Documentary Film*, 12 (2): 97–100.

Nguyen, K., Suied, C., Viaud-Delmon, I., & Warusfel, O. (2009) "Spatial audition in a static virtual environment: The role of auditory-visual interaction," *Journal of Virtual Reality and Broadcasting*, 6 (5).

Nozick, R. (1974) *Anarchy, State and Utopia.* New York: Basic Books.

O'Connor, M. B., Bennie, S. J., Deeks, H. M., Jamieson-Binnie, A., Jones, A. J., Shannon, R. J., Walters, R., Mitchell, T. J., Mulholland, A. J., & Glowacki, D. R. (2019) "Interactive molecular dynamics in virtual reality from quantum chemistry to drug binding: An open-source multi-person framework," *Journal of Chemical Physics*, 150 (22), [220901].

Pan, M. K. X. J., & Niemeyer, G. (2017) "Catching a real ball in virtual reality," *IEEE Virtual Reality (VR)*: 269–270.

Panofsky, E. (1934) "Jan van Eyck's Arnolfini portrait," *The Burlington Magazine for Connoisseurs*, 64 (372): 117–119.

Panofsky, E. (1947) "Style and medium in the motion pictures," *Critique: A Review of Contemporary Art*, 3: 5–28.

Patney, A., Salvi, M., Kim, J., Kaplanyan, A., Wyman, C., Benty, N., Luebke, D., & Lefohn, A. (2016) "Towards foveated rendering for gaze-tracked virtual reality," *ACM Transactions on Graphics (SIGGRAPH Asia)*, 35 (6): 1–12.

Peirce, C. S. (1974) *The Collected Works of Charles Sanders Peirce; Volumes V and VI*, edited by C. Hartshorne & P. Weiss. Cambridge, MA: Belknap Press.

Plantinga, C. (2005) "What a documentary is, after all," *The Journal of Aesthetics and Art Criticism*, 63 (2): 105–117.

Putnam, H. (1967) "Psychophysical predicates", in *Art, Mind, and Religion*, edited by W. Capitan & D. Merrill. Pittsburgh: University of Pittsburgh Press.

Putnam, H. (1988) *Representation and Reality*. Cambridge, MA: MIT Press.

Radford, C. (1975) "How can we be moved by the fate of Anna Karenina?" *Proceedings of the Aristotelian Society*, Supplemental Vol. 49: 67–80.

Regalado, A. (2020) "Elon Musk's neuralink is neuroscience theatre," *Technology Review*. Retrieved 16/02/2021 from https://www.technologyreview.com/2020/08/30/1007786/elon-musks-neuralink-demo-update-neuroscience-theater/.

Robson, J., & Meskin, A. (2016) "Video games as self-involving interactive fictions," *Journal of Aesthetics and Art Criticism*, 74(2): 165–177.

Scruton, R. (1983) "Photography and representation," in *The Aesthetic Understanding*. London: Methuen.

Searle, J. (1995) *The Construction of Social Reality*. Cambridge, MA: MIT Press.

Shehade, M., & Stylianou-Lambert, T. (2020) "Virtual reality in museums: Exploring the experiences of museum professionals," *Applied Science* 10, 4031; doi:10.3390/app10114031.

Silcox, M. (2017) *Experience Machines: The Philosophy of Virtual Worlds*. London: Rowman and Littlefield.

Slater, M. (2009) "Place illusion and plausibility can lead to realistic behaviour in immersive virtual environments," *Philosophical Transaction of the Royal Society London*, 364: 3549–3557.

Slater, M., Rovira, A., Southern, R., Swapp, D., Zhang, J. J., & Campbell, C., (2013) "Bystander responses to a violent incident in an immersive virtual environment," *PLoS ONE*, 8: e52766. doi:10.1371/journal.pone.0052766.

Slater, M., & Sanchez-Vives, M. V. (2016) "Enhancing our lives with immersive virtual reality," *Frontiers in Robotics and AI*, 3, 74.

Slater, M., Usoh, M., & Steed, A. (1995) "Taking steps: The influence of a walking technique on presence in virtual reality," *ACM Trans. on CHI, Special Issue on Virtual Reality Software and Technology*, 2 (3): 201–219.

Smith, M. A. (2001) "The Latin source of the fourteenth-century Italian translation of Alhacen's *De Aspectibus* (Vat. Lat. 4595)," *Arabic Sciences and Philosophy*, 11: 27–43.

Smith, S. (2018) "Dance performance and virtual reality: an investigation of current practice and a suggested tool for analysis," *International Journal of Performance Arts and Digital Media*, 14 (2): 199–214.

Smuts, A. (2009) "What is interactivity?" *The Journal of Aesthetic Education*, 43 (4): 53–73.

Spanlang, B., Fröhlich, T., Descalzo, F., Antley, A., & Slater, M. (2007) "*The making of a presence experiment: responses to virtual fire*," in PRESENCE 2007 – The 10th Annual International Workshop on Presence, Barcelona2007.

Steadman, P. (2001) *Vermeer's Camera: The Truth behind the Masterpieces*. Oxford: Oxford University Press.

Steinberg, L. (1953/1972) "The eye is part of the mind," reprinted in *Other Criteria: Confrontations with Twentieth Century Art*. Oxford: Oxford University Press.

Suits, B. (2014) *The Grasshopper: Games, Life and Utopia*. 3rd edn. Peterborough: Broadview Press.

Summers, N. (2020) "Why 'Second Life' developer Linden Lab gave up on its VR spin-off," retrieved 24/03/2021 from https://www.engadget.com/2020-03-27-why-second-life-linden-lab-sold-sansar.html.

Takatalo, J., Lehtonen, M., Häkkinen, J., Kaistinen, J., & Nyman, G. (2010) "Presence, involvement, and flow in digital games," in *Evaluating User Experiences in Games: Concept and Methods*, edited by R. Bernhaupt. London: Springer.

Tavinor, G. (2009) *The Art of Videogames*. Malden, MA: Wiley-Blackwell.

Tavinor, G. (2017) "*Fictionalism and videogame aggression*," Proceedings of DiGRA 2017, Melbourne, June 2017.

Tavinor, G. (2018) "Videogames and virtual media," in *The Aesthetics of Videogames*, edited by J. Robson and G. Tavinor. New York: Routledge.

Tavinor, G. (2019a) "On virtual transparency," *The Journal of Aesthetics and Art Criticism*, 77 (2): 145–156.

Tavinor, G. (2019b) "*Towards an analysis of virtual realism*," Proceedings of DiGRA 2019, Kyoto, August 2019.

Therrien, C. (2014) "Immersion," in *The Routledge Companion to Video Game Studies*, edited by M. J. P. Wolf and B. Perron. New York: Routledge.

Todorovic, D. (2009) "The effect of the observer vantage point on perceived distortions in linear perspective images," *Attention, Perception, & Psychophysics*, 71 (1): 183–193.

Toyota. (2019) "Why is Toyota developing humanoid robots?" Corporate Press Release, archived at https://global.toyota/en/newsroom/corporate/30609642.html.

Usoh, M., Arthur, K., Whitton, M., Bastos, R., Steed, A., Slater, M., & Brooks, F. (1999) "Walking > walking-in-place > flying in virtual environments," in *Proceedings of ACM SIGGRAPH 99*. ACM Press/ACM SIGGRAPH.

Van der Heijden, K., Rauschecker, J. P., de Gelder, B. et al. (2019) "Cortical mechanisms of spatial hearing," *Nature Reviews Neuroscience*, 20: 609–623.

Van der Meijden, O. A. J., & Schijven, M. P. (2009) "The value of haptic feedback in conventional and robot-assisted minimal invasive surgery and virtual reality training: A current review," *Surgical Endoscopy*, 23 (6): 1180–1190.

Walton, K. (1970) "Categories of art," *Philosophical Review*, 79 (3): 334–367.

Walton, K. (1978) "Fearing fictions," *Journal of Philosophy*, 75 (1): 5–27.

Walton, K. (1984) "Transparent pictures: On the nature of photographic realism," *Critical Inquiry*, 11 (2): 246–277.

Walton, K. (1990) *Mimesis as Make Believe*. Cambridge, MA: Cambridge University Press.

Walton, K. (2002) "Depiction, perception, and imagination: Responses to Wollheim," *The Journal of Aesthetics and Art Criticism*, 60: 27–35.

Welch, R. B., Blackmon, T. T., Liu, A., Mellers, B. A., & Stark, L. W. (1996) "The effects of pictorial realism, delay of visual feedback and observer interactivity on the subjective sense of presence," *Presence: Teleoperators and Virtual Environments*, 5 (3): 263–273.

Wildman, N., & Woodward, R. (2018) "Interactivity, fictionality, and incompleteness," in *The Aesthetics of Videogames*, edited by J. Robson and G. Tavinor. New York: Routledge.

Wilcox, L. M., & Allison, R. S. (2009) "Coarse-fine dichotomies in human stereopsis," *Vision Research*, 49: 2653–2665.

Witmer, B. G., & Singer, M. J. (1998) "Measuring presence in virtual environments: A Presence Questionnaire," *Presence*, 7(3): 225–240.

Wollheim, R. (1980) *Art and its Objects*, 2nd edn. Cambridge: Cambridge University Press.

Wollheim, R. (1987) *Painting as an Art*. Princeton, NJ: Princeton University Press.

Wollheim, R. (1998) "On pictorial representation," *The Journal of Aesthetics and Art Criticism*, 56 (3): 217–226.

Index

For Product Safety Concerns and Information please contact our EU
representative GPSR@taylorandfrancis.com
Taylor & Francis Verlag GmbH, Kaufingerstraße 24, 80331 München, Germany

www.ingramcontent.com/pod-product-compliance
Lightning Source LLC
Chambersburg PA
CBHW060449240326
41598CB00088B/4281

9 780367 620424